Studies in Fuzziness and Soft Computing

Volume 370

Series editor

Janusz Kacprzyk, Polish Academy of Sciences, Warsaw, Poland
e-mail: kacprzyk@ibspan.waw.pl

The series "Studies in Fuzziness and Soft Computing" contains publications on various topics in the area of soft computing, which include fuzzy sets, rough sets, neural networks, evolutionary computation, probabilistic and evidential reasoning, multi-valued logic, and related fields. The publications within "Studies in Fuzziness and Soft Computing" are primarily monographs and edited volumes. They cover significant recent developments in the field, both of a foundational and applicable character. An important feature of the series is its short publication time and world-wide distribution. This permits a rapid and broad dissemination of research results.

More information about this series at http://www.springer.com/series/2941

Jörg Verstraete

Artificial Intelligent Methods for Handling Spatial Data

Fuzzy Rulebase Systems and Gridded Data Problems

 Springer

Jörg Verstraete
Systems Research Institute
Polish Academy of Sciences
Warsaw, Poland

ISSN 1434-9922   ISSN 1860-0808   (electronic)
Studies in Fuzziness and Soft Computing
ISBN 978-3-030-13095-4   ISBN 978-3-030-00238-1   (eBook)
https://doi.org/10.1007/978-3-030-00238-1

This Springer imprint is published by the registered company Springer Nature Switzerland AG
The registered company address is: Gewerbestrasse 11, 6330 Cham, Switzerland

To Weronika, for her continuous support and enthusiasm

Foreword

The management and utilization of spatial data has become an important topic both for research and application. Data science attempts to make effective use of such data for numerous applications. In particular spatial data is now obtained in huge quantities from a variety of sensors and imaging sources. Much of the usage of spatial data requires it to be gridded but if data comes from a variety of sources, the grids are often incompatible. To resolve this problem typically a transformation is constructed by mapping a grid onto the other grids, which however introduces uncertainty and loss of accuracy. The research by Dr. Verstraete has been addressing these issues in his research for a number of years and provides a synopsis of many approaches here. However for data science this problem should be managed and interpreted in an intelligent fashion. In particular he looks at the fact that humans are able to bring their expertise and semantic understanding of domain specifics to obtain good regridding solutions. In order to develop this concept for the spatial problems such as regridding, spatial disaggregation, identification of locations, it is necessary to translate the spatial problem into a problem that can be solved using fuzzy rule- base systems. These capture the human knowledge and expertise to allow an automation of the solution to the spatial organization problems. These approaches are extensively demonstrated in several sets of experiments on disambiguation and regridding. Overall this volume is a significant contribution in the area of management issues required for the complex problems associated with applications involving spatial data.

Hancock County, USA
March 2018

Fred Petry
Stennis Space Center
US Naval Research Laboratory

Preface

This book combines and summarizes the author's research in developing an artificial intelligent method to cope with the map overlay problem for gridded datasets. This problem encompasses both spatial disaggregation, i.e. increasing the spatial resolution of a gridded dataset, and regridding, i.e. the remapping of a gridded dataset onto another raster; and both approaches take advantage of additional available proxy data to allow the artificial intelligent system to determine the underlying spatial distribution.

This book introduces the basic concepts and presents the issues related to overlaying gridded data in Part I. The shortcomings of the current approaches are indicated, and a novel concept is presented. Part II deals with the aspects of developing the concept into a theoretically sound system. This includes further developments of different aspects of rulebase systems, from the variables and ranges in the rulebase systems to the construction of the rules and further interpretation of the fuzzy results. Part III elaborates on different implementation aspects and demonstrates the performance of the developed method.

The results of this work are of interest to researchers and others who need to process gridded spatial datasets, while the development of the artificial intelligent method advances the state of the art of such systems. In particular, researchers working with fuzzy rulebase systems will discover the novel approaches in both rulebase construction and in the processing of the fuzzy outcomes of such systems.

Warsaw, Poland
March 2018

Jörg Verstraete

Acknowledgements

Acknowledgement has to go to Prof. Dr. hab. ir. Zbigniew Nahorski, of the Systems Research Institute, Polish Academy of Sciences, who not only introduced me to the map overlay problem but also supported looking at it from a different perspective. He also gave me the freedom to pursue alternative solutions, which allowed me to develop the presented methodology.

The Systems Research Institute, Polish Academy of Sciences supported my research and allowed me the freedom to follow my own direction. The colleagues at the institute provided helpful discussions and insights.

Contents

Part I
Preliminaries

The first part situates the research explained in the subsequent parts of this book. As such, it provides a general introduction into the field of research and introduces the concept of spatial data, as well as the structure of the book. The part then continues with the problem, its origin, and elaborates on related work. It also introduces the concept of using an artificial intelligent approach to solve spatial problems, without going into further technical details. This paves the path for explaining the necessary developments in Part II and the algorithm and experiments in Part III.

Chapter 1
Introduction

Abstract This chapter explains the current need for processing and combining data that carries a spatial component. A brief introduction first clarifies what exactly is data with a spatial component, before moving on to the conceptual spatial models that are commonly used in geographic information systems: feature based and grid based, along with their uses. The chapter elaborates on the details of the different models and why working with data in different representation models is unavoidable.

1.1 What is Spatial Data?

1.1.1 Introduction

In general, spatial data can be seen as a map on which different features are represented: the location of the features on the map (roads, cities, etc.) relates to their position in the real world and is the spatial aspect of this data. However, it does not have to be so visual: a set of coordinates (e.g. gps coordinates) also are spatial data, as they contain sufficient information to position the feature on a map. The concept of spatial data extends further than a single position and also includes spatial distributions of data, best illustrated by the weather forecast where different temperatures are shown over a region. Traditionally, spatial data is considered in two dimensions (2D), while some applications allow for 2.5D (2 dimensions with altitude information) and even 3D. Due to the spatial aspect of the data, the dimensionality of the data increases: for the example of the temperature, it is no longer sufficient to consider a set of values (one dimension), but the location of the data needs to be considered. This requires additional care as the spatial aspect cannot be lost. First, there is an issue of representation, and due to the multitude of types of spatial data, there are different ways of representing spatial data (Sect. 1.2): representing the position of an object or person is different from representing the distributions of temperatures of a region, yet both types of data can occur. To complicate things, these different representations need to be interoperable: if the position of a person is known, and the distribution of temperature is known, it makes sense to ask a system what the temperature is at that person's location. Another important aspect relates to the accuracy and granularity of the data: temperature for example has a value for every square centimeter, but of course it is impossible to gather and store data at such an accuracy. But what is an acceptable accuracy? Is a single temperature value sufficient for every $100\,\text{km}^2$? How accurate and representative is this value? The answer to this depends on the type of data that is modelled, the needs, possibilities for gathering the data and the infrastructure for storing the data. All these issues combined make the processing and in particular the combination of spatial data from different sources a non-trivial task as the questions are answered differently for different data.

Over the years, a lot of progress was made in remote sensing and modelling, allowing now for huge amounts of data to be collected or simulated. The way data are represented is closely related to how the data were obtained or calculated. Many researchers often need to work with different types of data; this usually implies that the data is obtained from different sources. As a result, there is a need for combining differently modelled data, which is a non-trivial problem. While the techniques for data gathering are improving and increasing the accuracy of the spatial data, both regarding the resolution and the contained data, this also increases the expectations of what knowledge can be extracted from the data. As more datasets on different topics are gathered, the need and desire to combine them increases.

A lot of today's research involves data that contains spatial data. This means that the data carries a spatial component, associating the information with a location or region. Examples are climate research, where data such as temperature or humidity not only concern the measured values or the values obtained through models, but also the locations at which these measurements were made. Another example concerns air pollution [4, 16, 20, 33], where data on pollutants is also obtained through models and measurements, but additional information can involve the location of both point sources (e.g. factories), line sources (e.g. roads) or areas (e.g. cities). Another relevant example is in epidemiology, where the conditions in which certain bacteria thrive need to be combined in order to identify risk areas. For example, Vibrio Cholerae, the bacteria that causes cholera, thrives in water with low salinity in areas with a high temperature. Research in identifying risk-areas [25] thus needs data on air temperature to be combined with salinity levels of nearby water, information that in turn has be connected to populated areas. Even determining the salinity level may require the use of different data, as e.g. data on rainfall can be used to asses salinity levels. Other fields that often need to use spatial data are archaeology or assessing the risk and effect of natural disasters.

The real world example that triggered our research was related to assessing the exposure to air pollution where population data has to be compared against both measured data regarding pollutants and dispersion models to estimate presence of pollutants. The focus in the research presented here is to combine spatial data, independent of source or meaning of the data, using artificial intelligence. The main reason for opting for artificial intelligence lies in the complexity of the problems, in part due to the higher dimensionality caused by the spatial aspects of the data, but also due to the complex connections that can occur between the data. This research combines aspects from spatial data with artificial intelligence. To provide insight in the different chapters and sections, the structure of the book is explained in the next section.

1.1.2 Organization of This Book

This book summarizes the results of the development of a fuzzy rulebase system for spatial data processing. This approach was initially developed to solve the problem of spatial disaggregation, but is general enough to be extended to the general map overlay problem for gridded data and multiple additional applications. Before introducing the concept, it is first necessary to elaborate on spatial data models, not only to understand the issues at hand, but also to grasp the complexity of the problems and the difficulties in combining the data.

Continued in this chapter is an introduction to the models for representing spatial data (Sect. 1.2); followed by a short explanation on how these models are used and how data are processed (Sect. 1.2.5), paying particular attention to gridded models, as they are at the center of this research. Chapter 2 deals with the problems that occur when gridded datasets are combined and presents current solution methods; the concept of the developed approach, and the mimicked intelligent reasoning are explained in Chap. 3.

Part II turns to the development of the concept and the necessary advancements to the state of the art in order to develop the methodology. Chapter 4 introduces the known concept of fuzzy rulebase systems, which will be adopted to process the spatial data. The subsequent chapters elaborate on specific developments needed to translate the spatial problem. This starts with Chap. 5, where attention goes to how to define parameters and most possible ranges for the spatial problem. The ability to calculate specific most possible ranges allows for a more efficient rulebase construction method, which is developed in Chap. 6. The application of the fuzzy rulebase system allows for additional constraints that can be used to better interpret the fuzzy sets output by the fuzzy rulebase system. For this purpose, a new defuzzification method that allows for defuzzifying multiple fuzzy sets under a shared constraint is developed in Chap. 7. While not directly related to the development of the fuzzy rulebase system, comparing gridded datasets is necessary for both evaluating the method but also within the creation of the fuzzy rulebase system to determine the most suitable parameters. Chapter 8 explains the shortcomings in the current approaches to compare datasets when applied in the context of spatial data and introduces a newly developed method that compares datasets by judging how well they seem to approximate an unknowing underlying spatial distribution of the data. The last chapter of this part, Chap. 9 summarizes the algorithm, combining the developments described in the earlier chapters.

In Part III, the algorithm is further analysed (Chap. 10), followed by a presentation on the practical implementation aspects and issues in Chap. 10. In Chap. 11, the performance of the developed method for spatial data processing is verified and studied using experiments.

1.2   Spatial Data Models

1.2.1   Overview

The term spatial data covers a wide range of aspects, and these need special attention and matching tools. To work with spatial data in computer systems, Geographic
Information Systems or GIS for short, were developed. These are software packages
specifically designed to allow a user to handle spatial information and as such they
have provisions for, among others, entering, editing, querying and displaying geographic data. Data that carries a spatial dimension can be represented internally in
different ways, depending on the application and use. In this section the two most
common models used in GIS are presented: feature based models and grid based
models; these are two entirely different models that are used to connect data with
locations, with very distinct applications [34, 35]. Both have their specific applications and uses, and GIS systems are capable of handling both of them. For geographic
data, apart from different possibilities of representing the data, there additionally is
the problem of mapping the data onto a flat surface, and many projections and coordinate systems exist along with associated problems related to the conversion between
them. It is important to realize that in this section, and in general this book, the
problems are considered at the level of the models; we are not concerned with geo-
referencing or projections. In this work, all data are considered to be geo-referenced,
implying that the available data is already considered to be in the same projection
and the same coordinate system. The topic of this work is not in performing spatial
transformations to perform this geo-referencing, but in analyzing and relating data
that is already geo-referenced. Even when data are geo-referenced, combining data
is still not a trivial task. First, it is necessary to explain the data models used in the
GIS. At model level, there are two main categories that can be distinguished: feature-
based models and grid-based models; this work concerns the combination of data
represented in a grid-based model. The subsequent sections provide more details on
both these models, as both will be referenced in this book.

1.2.2   Feature Based Models

The feature based models in geographic information systems are the most intuitive.
Real world objects, in GIS terms called *features*, are represented by simple geometries: points for locations, lines for borders or boundaries and polygons for areas. With
each feature, additional attributes can be associated, these attributes can be names
or descriptions, but also numeric data: a point could represent a measuring station to
measure the amount an air pollutant or a factory as a source of a pollutant. Both of
these could have some additional data such as a name but also numeric information
that would be the measured amount of that pollutant. A line could represent a road,
with a name, possibly a category of the road, but also with for example the average

Fig. 1.1 Example of the feature based models, where points, lines and polygons match real world features. This particular example is a section of Warsaw, showing roads (line segments), city squares and the Vistula (polygons)

number of cars per day. In feature-based models, there is an aspect of scale and not just at visual level: when viewing an entire country, a city could be represented by a point, however at a smaller scale the city would rather be represented by a polygon showing its administrative outline. As the basic element is a straight line segment, this outline may need to be more refined as the scale increases. Roads can be represented as lines, however at smaller scales they are represented by multiple lines to indicate the sides of the road or even multiple lanes. As such, the scale at which the data will be viewed and used may affect the way the features are represented. In the example on Fig. 1.1, the roads are represented by single line segments, making it impossible from these geometries to judge the size of the roads; however information on this can be contained in associated attributes. In this dataset, each road segment has data associated that contains the name of the road, but also the information on the type of road.

Feature based models are important as they allow a highly accurate representation of the environment, even valid at almost any scale if the features are modelled with enough detail. As the detail can depend on the application, these models are also quite efficient regarding storage in a computer system, only requiring the vertices needed to define the points, lines or polygons at the desired precision. The feature based models are commonly used for representing maps that model real world objects (e.g. roads) or concepts (such as administrative borders). The GIS offers specific operations for working with features; this include operations that derive data from the geometries (e.g. surface area, circumference or distance) but also topological operations to determine who two given geometries relate to one another and operations to calculate e.g. the intersection of given geometries.

1.2.3   Grid Based Models

Grid-based models, also called raster based models, serve an entirely different purpose from feature based models: they allow the representation of a specific (usually numerical) property over a region of interest. The modelled property has values that differ with the location, and as such has a spatial distribution. An examples of this are the temperature or the amount of a pollutant: there is a value for every location in the region of interest (ROI), and this value differs per location. Grid-based models allow the representation of such data by discretizing the region of interest, thus limiting the amount of data. While the property quite often has a continuous change in value (e.g. temperature), in practise it is impossible to collect the data as such, resulting in available data that is not continuous over the region. The grid itself partitions the region of interest, which means that the region of interest is divided in a number of non-overlapping cells that cover it completely. Typically, rectangular grid cells are used (depending on the coordinate system and projection used to visualise them), but hexagonal cells or even an irregular partitioning are possible alternatives. Irregular partitioning can be useful or necessary when for example administrative boundaries are used. Every cell of the grid is assigned a value of the modelled property; this value is deemed representative for the covered area; this can be an average (e.g. in the case of temperature), a total (e.g. for the amount of a pollutant) or any other aggregation that is suitable for the application.

A grid is said to have a (spatial) resolution, on Fig. 1.2 an example is given of a grid that covers Warsaw, where each cell covers an area of 2 km × 2 km. The size of the grid cells determines the scale at which the data can be distinguished: cells of

Fig. 1.2 Example of a grid-based model, overlayed with the above featurebased model. With each grid cell, several values are associated, and two of those are indicated using the bar-charts centered in the grid cells

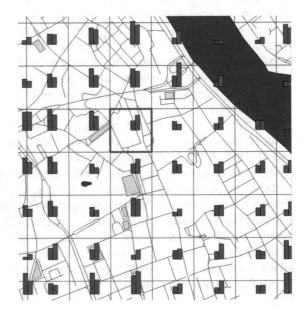

Fig. 1.3 Example of possible underlying distributions: the grid is unable to distinguish whether data associated with a cell originates from specific features, each of the shown underlying distributions (single point, two points, line, areas of various sizes) can yield a similar grid cell

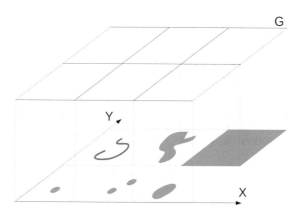

1 km × 1 km can for example hold data that is more spatially accurate, but obtaining data at a higher resolution may be more difficult or impossible. It is important to realize that the grid contains no information regarding a further underlying distribution of the underlying data: the grid cells are the smallest possible unit considered. Data is associated with the grid in *bands*, where one band holds all data of the same type (e.g. temperature). The grid can have multiple bands associated. Operations on grid based models are more limited than those on feature-based models, as no geometries are involved and mainly tend to concern calculations involving the associated data.

While the grid itself provides no information regarding the spatial distribution of the data inside a grid cell, the grid is usually an approximation of a real world situation and as such there can be different underlying distributions. As an example, consider the grid on Fig. 1.3. If we consider this an example of sources of air pollution, then a point source can represent e.g. a factory, and a line source can represent e.g. a road. While both can be sources of a pollutant that has a spatial component, the gridcells of the grid that represents the sources is unable to distinguish between these different types. This lack of knowledge regarding the underlying distributions complicates processing grids. Quite often, a uniform distribution is assumed as this is the easiest assumption and lacking more information on the modelled it is a natural and intuitive choice. However, if the reality deviates from this spatial distribution, subsequent calculations and analysis will be prone to uncertainty and imprecision.

1.2.4   Digital Elevation Models

Somewhat related to the grid based models are the digital elevation models. These are also used to model an attribute whose values changes in space, but use a different technique to approximate the spatial distribution. Commonly, these models used Delauney triangulation [35] to span a network of triangles over a set of vertices, as shown on Fig. 1.4. The Delauney triangulation is an algorithm that generates

Fig. 1.4 Delauney triangles
are commonly used in digital
elevation models. Here, eight
points are sampled of the
fluid shape and the triangular
network is built on these
eight points. It is clear from
the result that eight is
perhaps too little to properly
approximate the original
shape

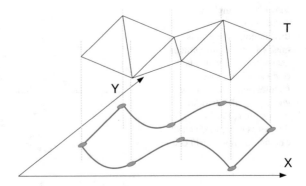

the network of triangles in such a way that most triangles resemble the equilateral triangle [36], thus avoiding very long, narrow triangles; the triangles are the duals of Voronoi diagrams. For a given set of vertices, this algorithm result in a unique definition of the triangles. The numerical data to be modelled is associated with the vertices: these represent the points where data is available. This numerical data can then be interpreted as a third dimension, causing the triangles to form three dimensional planes in space. These planes are then used as a means of linear interpolation: the value of any point inside a triangle or on an edge of the triangle is obtained by determining the value of this third dimension. Due to the fact that the structures are triangles, this is a quite straightforward operation using linear interpolation.

This model is often used for altitude models, as this data is quite static and it is possible to obtain measurements on many places. For very flat regions, or regions where the linear interpolation closely resembles the reality, not many vertices are needed to have an accurate representation of the reality. On the other hand, for regions with many differences in elevation, vertices can be located much closer together. This allows the triangular model to be quite optimal for storing altitudes over an area. One important issue is that the definition of the triangles can change when a vertex is added or removed, and although these changes are localized around the added or removed point, it alters the interpolation and thus affects the how the values are calculated. This is why this model is more used for static data that can be verified and for which the vertices can be optimized to have the most suitable interpolation.

1.2.5   Data Structuring in GIS

A geographic information system represents different data in layers; a layer typically holds one type of data. User can then specify to view specific layers and extract information from different layers at once, allowing them to analyse a multitude of data, from different sources. The GIS takes care of coordinate transformations, using the information of each layer regarding their associated coordinate system and projection, and is capable of geo-referencing the data and displaying them using the system and projection specified by the user. In the Open Geospatial standard (formerly OpenGIS, [32]), and libraries such as JTS [21] that follow the standard, a layer is limited to data of the same type: for a feature based dataset, this means that the layer can thus contain points, or lines, or polygons but not combined. A dataset on e.g. water can therefore consist of one layer with lines (e.g. to represent the narrow rivers) and one with polygons (e.g. to represent lakes or bigger areas).

Each layer has a number of attributes associated, with a name and a datatype, for which the contained dataset may or may not have values (for a gridded dataset, these attributes match the bands). Apart from allowing the geometries of the contained features to be manipulated, the GIS system also provides access to the attribute table of the layer, allowing modifications, selection and (limited) analysis of the attribute data.

The GIS geo-references all the available information and presents this to the user. Commonly, an end user requires information from different datasets and aims to compare, correlate or combine the available data. This operation is called a map overlay: the data of one dataset is overlayed with the other dataset. The capabilities of the GIS for geo-referencing provides the end user with a visual guidance, however with datasets originating from different sources and containing different types information, combining the actual data contained in the datasets is a not a trivial matter. It may be necessary to combine two or more feature-based datasets, two or more gridded datasets or a combination of both. The GIS provides operations at geometric level, to identify e.g. which features of one layer overlap with features of another layer, determine the intersection or various other operations, also including concepts such as bounding rectangles and buffers.

While combining feature based data may seem straightforward at first, one has to take into account the possible issues of the scale differences. As mentioned in Sect. 1.2.2, the detail level of a feature depends on the application: for some applications, representing cities as single points is sufficient. Combining such a dataset with a dataset that represents the highways is more or less possible, but combining it with a dataset that holds a street network is problematic and cannot be solved without more information on the outline of the city.

For gridded datasets, the issues are quite different. Gridded datasets are often specified in a format such as GeoTIFF, which is best described as a geo-referenced bitmap (conceptually not unlike a digital photo, even though the content not necessarily has to be an image). The pixels of the bitmap represent the cells of the grid and with each cell one or more bands (this continues with the analogy with image bitmaps, where multiple bands are present for colours, transparency or other information) which hold numeric data. Sometimes, in order to be able to perform specific operation, another grid representation is used: the grid is comprised of a number of geometries, each of these representing one grid cell. While essentially the same type of model, the internal representation is quite different and facilitates specific operations: it is for example easier to determine how much a cell in one grid overlaps with a cell in the other grid, as it suffices the calculate the intersection of the features involved. In this work, the focus will be on gridded data; in order to perform the desired operations, this latter representation is used.

Chapter 2 specifically elaborates on the difficulties and issues related to overlaying gridded datasets. The chapter explains why the lack of knowledge of the underlying distribution complicates the operation and details current ways in which the map overlay is performed. The concept of the solution method put forward in this book is described in Chap. 3.

Chapter 2
Problem Description and Related Work

Abstract This chapter explains the problem with overlaying gridded data. It explains why there is a need to remap an existing grid onto a different grid (a procedure called regridding) and elaborates on the details of the specific case of spatial disaggregation as well as the general case of regridding ill-aligned grids. Following the introduction of the problems related to overlaying spatial data, particularly in the context of gridded data, it elaborates on related work and state of the art, and continues with the formulation of the problem and the concept of the solution considered in this work.

2.1 Overlaying Gridded Data

The map overlay operation relates to combining or comparing data in different layers. Overlaying gridded data implies correlating the information in cells from one grid to the cells in the other grid. If both datasets are defined on the same raster, the grids are called compatible: all the cells line up and comparing data of both grids is very straight forward, as shown on Fig. 2.1.

Performing the map overlay on gridded data is often described by means of Tomlin's map algebra [39]. Tomlin's map algebra provides definitions for operations to determine a value of an output cell using data in multiple grids, classified based on which cells are considered. Local operations only use cells from other grids that are in the same location as the output cell and usually use simple arithmetic operations: the value for a cell is e.g. the average of the values of the cells in the same location of the different grids. Zonal operations use cells from a neighbourhood defined around the cell of interest, combining the values of all cells in this neighbourhood. Global operations, lastly, use the entire grid to determine values. The application of Tomlin's map algebra requires the grids to be defined similarly and assumes there is a one-one mapping between the cells of both grids, as shown on Fig. 2.1.

However, when data comes from different sources, the grids are more likely to be incompatible. This means that the rasters do not not line up, as illustrated on Fig. 2.2. This can happen when the grids are defined on a different position; use a different resolution, e.g. grid G_1 has 1.5 km × 1.5 km grid cells, whereas grid G_2 has 2 km × 2 km grid cells (Fig. 2.2a); when the grids are rotated relative to one another

© Springer Nature Switzerland AG 2019
J. Verstraete, *Artificial Intelligent Methods for Handling Spatial Data*, Studies in Fuzziness and Soft Computing 370, https://doi.org/10.1007/978-3-030-00238-1_2

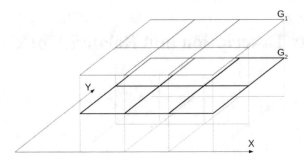

Fig. 2.1 Example of compatible grids. The values of cells in G_1 can easily compared with the values of the cells in G_2 as the cells cover the same area

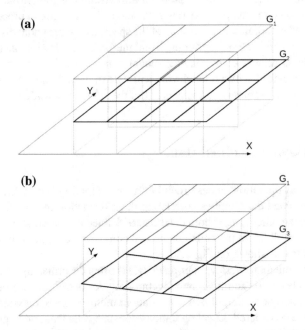

Fig. 2.2 Different examples of incompatible grids, in (**a**) the grid-size differs whereas in (**b**) the grids are rotated compared to one another

(Fig. 2.2b); when the definition of a grids has a shifted position compared to the other grid, or any combination of these.

In order to be able to relate grids that are not compatible, one grid will have to be remapped onto the other grid, to make the grids compatible. This means that the data of the cells in one grid have to be redistributed to match the grid layout of the other grid. This will result in a one-to-one-mapping between cells of both grids, allowing them to be matched, and allowing Tomlin's map algebra to be applied. The problem of remapping a grid onto a different grid is called *regridding*. Regridding is not a trivial task, as the grid contains no information concerning the underlying spatial distribution of the data in each grid cell.

As the development of the artificial intelligent approach to regridding is the core topic of this book, the next sections explain the associated problems (Sect. 2.2) and the current solution methods (Sect. 2.3). This highlights the shortcomings of the current approaches and paves the way for introducing the concept of the developed system in Chap. 3.

2.2 Overlaying Ill-Aligned Grids

2.2.1 Introduction

Ill-aligned grids, or incompatible grids are grids where there is no one-to-one mapping between the grid cells. Both grids are geo-referenced and cover the same region of interest (not necessarily to the same extent), as shown on Fig. 2.2, but their grid cells do no overlap nicely. This can happen if the cells have different sizes (one grid could have cells that are 2 km × 2 km, another could have cells that are 1.5 km × 1.5 km), the cells could have the same sizes but the grids are shifted towards one another, on grid can be rotated compared to the other grid, or any combination of these. The underlying reason for this difference is related to the data that is modelled: there can different remote sensing methods, different measurements, different standards, etc., all of which are possible causes for having to deal with incompatible grids.

2.2.2 Regridding

The typical approach to regridding is to map one grid onto the other grid. The choice of which grid layout will be used depends on a number of factors, but will always result in losses of accuracy and certainty. One can choose the most fine grid (the one with the highest resolution) to try and limit the losses in accuracy, or opt for the least fine grid to try to limit the uncertainty.

For describing regridding, it is assumed that the grid represents an absolute value that is cumulative, such as *total population* or *amount of ppm of a pollutant*. If by contrast the grid would represent values that are relative to the area, for example *amount of ppm per square kilometre* or *people per square kilometre*, the area aspect has to be eliminated from the property prior to considering this constraint. This is a simple operation, as conversion to absolute value is achieved by multiplying the value with the considered area. After the regridding operation, the inverse operation can be applied to return to the relative property. This is a necessary step prior to performing the remapping; in the subsequent discussions on remapping throughout this book,

Fig. 2.3 Example used to
illustrate the remapping of
gridded data: grid A will be
mapped onto grid B

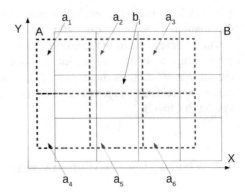

all grids are considered to contain absolute values. The reason for this assumption
is that it imposes an easy to verify and intuitive constraint, as all cells should sum
up to the same value. For some attributes, such as temperature, a different constraint
needs to be used. In case of temperature, the requirement that the area-weighted
average should be constant is a candidate. While the developed techniques should
be applicable also for such data, the simpler and intuitive constraints with absolute
value data facilitate describing and reasoning about the regridding problem.

Consider two grids A and B as shown on Fig. 2.3. The notation $f(\cdot)$ is used as
a notation for the value associated with a given cell. Remapping the grid A to the
different grid B is equivalent to finding weights x_i^j, so that the value $f(b_i)$ of a cell
b_i in the output grid B is the weighted sum of the values $f(a_i)$ of the cells a_i in the
input grid A.

$$f(b_i) = \sum_j x_i^j f(a_j) \tag{2.1}$$

$$= \sum_{j|a_j \cap b_i \neq \emptyset} x_i^j f(a_j) \tag{2.2}$$

As a cell b_i intersects with a set of cells of A, it makes sense that its value $f(b_i)$ is
calculated using only those cells, hence only cells of grid A that intersect with the
target cell b_i will contribute to $f(b_i)$. The formula (2.1) can therefore be modified
to only consider these intersecting cells (2.2). The weights x_i^j in the above formula
are thus subject to the following constraints:

$$\forall j, \forall i, x_i^j \geq 0 \tag{2.3}$$

$$\forall j, \sum_{i|a_j \cap b_i \neq \emptyset} x_i^j = 1 \tag{2.4}$$

Constraint (2.3) states that the weights are positive. This is because both grids are
approximating the same underlying distribution, high values in one grid should relate

Fig. 2.4 Illustration of the spatial disaggregation: grid A needs to be mapped onto grid BB, a grid in which each cell is a part of a cell of grid A

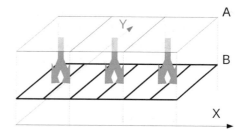

to high values in the other. As such, values of cells in grid A cannot negatively affect values of cells in grid B. Both grids are an approximation of a real quantifiable property, whose distribution should not change by changing the representation. Constraint (2.4) guarantees that the entire value of an input cell is remapped, but also not more than the entire value; as both grids represent they same property, their total value should be the same ($\sum_i f(b_i) = \sum_j f(a_j)$). The combination of both constraints implies that $x_i^j \leq 1$, for all i, j.

2.2.3 Spatial Disaggregation

A special case of the general regridding problem is spatial disaggregation. This occurs when grid B is such that its cells partition the cells of the other grid A; this implies that the data of every cell of A has to be distributed over an exact number of cells of B as illustrated for example in Fig. 2.4.

It is called spatial disaggregation, as each cell of A can be considered an aggregation of a number of cells in B. As every cell of B is a portion of a cell in A, the formula that defines values of cells b_i of grid B is:

$$f(b_i) = x_i^j f(a_j) \tag{2.5}$$

The constraints on the weights x_i^j in the above formula are subject to the following similar constraints from the general grid remapping in Eqs. (2.3) and (2.4).

However, due to the specific relative layout between both grids, the following property also holds:

$$\forall j, \sum_{i | a_j \cap b_i \neq \emptyset} f(b_i) = f(a_j) \tag{2.6}$$

This constraint is stronger than the second constraint in the general remapping (Eq. 2.4), and also stronger than the global constraint that all cells of both grids should sum up to the same value. It turns the problem into a very localized problem: the disaggregation of a cell is fully independent of the disaggregation of neighbouring cells, as there is no partial overlap. While for some matters this is a simplified problem, the lack of partly overlapping cells also provides less information on how the data of the cells should be remapped. While the origin of the problem is similar, the specifics of the spatial disaggregation allows for specialized methods.

2.3 Related Work

2.3.1 Areal Weighting

Areal weighting [13] is the simplest and most commonly used approach for regridding; it uses the area of the overlap of a cell a_j in grid A and a cell b_i in grid B to determine the portion of a_j that will be mapped into cell b_i. The weights are calculated as the portion of overlap between the input cell and the output cell: this portion is the surface area of the intersection over the surface area of the input cell.

$$x_i^j = \frac{S(A_j \cap B_i)}{S(A_j)} \tag{2.7}$$

The conditions for general remapping are still met:

$$f(B_i) = \sum_{j|A_j \cap B_i \neq \emptyset} x_i^j f(A_j) \tag{2.8}$$

$$= \sum_{j|A_j \cap B_i \neq \emptyset} \frac{S(A_j \cap B_i)}{S(A_j)} f(A_j) \tag{2.9}$$

The areal weighting method is very simple in calculation and therefore the most commonly used. The assumption regarding the spatial distribution of the underlying data is that it is uniform inside each cell and independent of other (neighbouring) cells. Barring other information, this is a fair assumption, but it often is too much of a simplification. If the value of a cell stems from localized sources (point or line sources) in a rather small part, this method performs quite poorly, yielding a less accurate result where the data are more spread out than in the initial grid. The method is also poorly suited for spatial disaggregation as, in this case, each cell b_i of grid B would get the same value (Fig. 2.5).

2.3.2 Spatial Smoothing

In spatial smoothing, the assumption on the underlying distribution is different than in areal weighting. This time, the data are assumed to have a smooth distribution over the

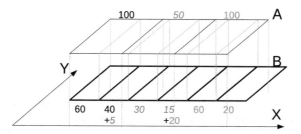

Fig. 2.5 Illustration of the areal weighting: the cells of grid A are mapped onto grid B using the amount of overlap as weights (values are illustrative)

entire grid [14, 38]. The modelled data can be envisioned as a third dimension, which makes the grid resemble a stepped surface. The assumption of a smooth distribution over the entire grid is achieved by fitting a smooth surface over this stepped surface, such that the volume under the smooth surface is the same as the total volume in each of the grid cells. Once this surface is determined, the regridding is achieved by sampling the surface using the target grid; each cell is assigned the value of the volume it matches with. This method outperforms areal weighting when the underlying is indeed smooth. However, the assumption that the spatial distribution is smooth does not always match the real situation. Consider for example two neighbouring cells, each of which has its value determined by a single point source and all other neighbouring cells have a value of 0. Assuming a smooth distribution may result in the smooth surface to have a highest point or ridge, in between these two cells. This would be fine if both points would be close together, but not if both points were on opposite sides. On Fig. 2.6, spatial smoothing is illustrated.

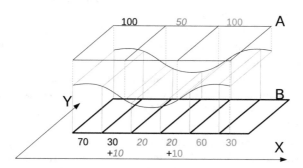

Fig. 2.6 Illustration of spatial smoothing: a smooth surface is derived from grid A, grid B is determined by resampling this surface (values are illustrative)

2.3.3   Spatial Regression

In regression methods, statistics are used to establish patterns in the overlap of the
data [10]. Different variations exist based on they way these patterns were established.
In [11], the authors determine zones. Using these zones and an assumption of the
distribution of the data (e.g. Poisson), it is possible to determine value for incompatible areas; these are areas with grid cells that do not have a one-one overlap. For this
purpose, there is a choice of underlying theoretical models, but each of them requires
making key assumptions that are not part of the data and cannot be verified using the
data at hand. These assumptions concern the distribution of the data. Consequently,
if these assumptions do not match the real underlying distribution, the regression
method also can yield worse results. The authors in [27, 28] employ Markov-chain
Monte Carlo computational methods to produce estimates. In [17], the authors aim
to use additional information, from additional geographic datasets (so called proxy
data) to try and determine a fitting regression model.

2.3.4   Data Fusion

Data fusion is a different problem from grid remapping, but it is somewhat related
to the method presented in this book. The aim in data fusion methods is to combine different datasets that contain the same property in order derive a single dataset
of either higher quality or with more detail. This differs from the proposed regridding process where the primary aim is to use data containing different properties to
improve the regridding process. Contrary to the datafusion techniques, the data used
is not necessarily similar and the goal is not to generate a combined dataset but to
achieve a higher quality remapping of a given dataset.

The authors in [44] present a statistical approach to combine different datasets
that contain the same data. Their work fits in the feature based model (Sect. 1.2.2),
as they aim to combine multiple sources of geometries that should relate to the
same information in order to derive a single dataset with higher accuracy. Their
methodology is applied on an example of the German road network, combining two
different sources (EDRA and EGT) for data on the road network, but both sets differ
both geometrically and topologically.

In [12], the authors combine datasets that contain information on vegetation. They
apply data fusion not to improve the definition of the geometries of the features, but
rather to improve the data associated with the features. This information in their
case is not numerical and is not presented in a gridded format, but rather it is in
the form of annotations to regions on the map (also a feature based approach). The
different source data contains information on the same topic, but the annotations
use different definitions and vocabulary while also the regions in both maps differ.
The authors' approach consists of first building a common ontology and identifying

possible overlaps and conflicts. The ontology is then used to determine which term will be used for this particular region in the end result.

While a different problem than spatial disaggregation or regridding in general, data fusion is mentioned here as there are similarities to the research elaborated further on in this book. The developed methodology makes use of additional data, and while this is a novel approach to regridding, the successes in data fusion show that combining data can have its merits. In addition, the developed artificial intelligent method for grid remapping can also be extended to perform data fusion.

Chapter 3
Concept

Abstract The concept of the methodology developed in this book is described in this chapter. It approaches the problem of regridding by looking differently at it: from the point of an expert who has knowledge on the topic and how he/she would use that knowledge to obtain better results. A short introduction to a rulebase system is provided—this will be explained in more detail more in Part II, particularly in Chap. 4.

3.1 A Different Look at the Problem

3.1.1 Spatial Disaggregation

The problems of regridding and spatial disaggregation were explained in Sect. 2.2, along with current solution methods. Without any further knowledge, any of the possible solution methods is equally good—or equally bad—as long as the assumptions are within reason. The situation changes if more would be known about the underlying distribution: an expert with knowledge on the topic can make educated assumptions on the underlying spatial distribution and perform a better remapping of the data. As it turns out, quite often, more is known about the underlying spatial distribution than is immediately apparent: the underlying spatial distribution of the property may be related to the spatial distribution of other known datasets. As such, it should be possible to use these data sets as so called *proxy* data to improve the regridding or disaggregation. The example that inspired the automation of the thought process of performing this remapping stems from air pollution: it should be possible to perform a better spatial disaggregation of a gridded dataset that represents the concentration of an air pollutant that relates to car traffic using information on the road network and traffic data as proxy data. Incorporating the proxy data is not straightforward, as the data may not correlate exactly and given proxy data may not be the only explanation for the underlying distribution. As such, the proxy data should not be followed too strictly. To achieve this, we first consider how an expert can reason about this problem using proxy data.

© Springer Nature Switzerland AG 2019
J. Verstraete, *Artificial Intelligent Methods for Handling Spatial Data*, Studies in Fuzziness and Soft Computing 370, https://doi.org/10.1007/978-3-030-00238-1_3

Fig. 3.1 Illustration of the
spatial disaggregation: grid
A needs to be mapped onto
grid B; proxy data C
provides information on the
underlying distribution

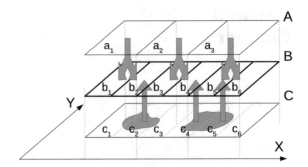

Consider a grid A of input data, a grid B onto which the data should be regridded
and a proxy grid C as illustrated on Fig. 3.1. Using the same notation as in Sect. 2.2,
the value of a cell x_i of a grid X is denoted $f(x_i)$.

Performing the regridding of grid A onto grid B using the information that the
data also relates to grid C is not necessarily possible when interpreting this relation
strictly, as the data may not match perfectly (the proxy data from a different source
and may not reflect the relation accurately). The remapping should therefore follow
the spatial distribution of C rather loosely. If we consider $f(a_1)$ the value of a_1,
then its value should be distributed over b_1 and b_2. At the same time, a_1 overlaps
with c_1, c_2 and even a part of c_3. Similarly, b_1 overlaps with c_1 while b_2 overlaps
with c_2. Next, let us assume that C correlates positively with A, so that high values
of C coincide with high values in A. If the value $f(c_1)$ is greater than the value
$f(c_2)$, then the connection with grid A implies that the remapping to grid B will
result in $f(b_1) > f(b_2)$. Similarly, if $f(c_2)$ would be greater than $f(c_1)$, then it
implies that $f(b_2) > f(b_1)$. A similar reasoning can be performed for all possible
combinations; the problem is how to translate this reasoning. On Fig. 3.1, if C is a
grid that approximates the underlying distribution indicated by the shaded area, this
reasoning will result in a better disaggregation of grid A. This example is simplified,
as grid C is defined on the same raster as grid B; this is however not a requirement.
Moreover, the proxy data contained in C does not even have to be gridded data; it
can even be a feature based dataset that contains data providing information on the
underlying spatial distribution of the data that is regridded. More formally, an expert
uses a reasoning using rules of the form

```
IF [values in C in the area of b_i are high]
   THEN [the portion of A mapped to f(b_i)] is high
IF [values in C in the area of b_i are low]
   THEN [the portion of A mapped to f(b_i)] is low
...
```

These rules leave a number of aspects still open for interpretation: What are
values in C? What is *in the area of*? The first question relates to which data of C
is considered, and indicates the data associated with the different cells. However, an
expert could also consider more complicated predicates, not just considering values
in C, but actually considering values derived using the data in C. The second question

concerns which cells of C are considered to determine the portion to be mapped from a given cell a_i to its overlapping cells in B. Both these questions can be grouped by rewriting predicate in a more general way, in way that considers calculating specific properties from C, which is more general than the positive connection with A in overlapping areas.

```
IF [values calculated from C in function of b_i are high]
   THEN [the portion of A mapped to f(b_i)] is high
IF [values calculated from C in function of b_i are low]
   THEN [the portion of A mapped to f(b_i)] is low
. . .
```

The key in making this approach work, is to correctly translate the problem into a formulation that fits with these rules. This will allow us to solve the problem by adapting a fuzzy rulebase system, a known technique in artificial intelligence. To achieve this, it is necessary to determine how to calculate values from grid C and how to relate them to grid A. It is also necessary to define the linguistic terms against which these values will be evaluated, these are the terms *low* and *high* in the above rules. This not only includes defining how to represent low and high, but also defining the domain in which these linguistic terms are defined. Similarly, the linguistic terms and domains for the consequence need to be defined. This will be explained in Chap. 5 in Part II.

3.1.2 Regridding

The problem of regridding differs from spatial disaggregation in the fact that there can be partial overlap of cells in the input and cells in the target. This changes how data can be remapped, as now the problem is not localized: in spatial dissaggregation, every input cell can be dissaggregated independently of the other input cells; the value of an output cell is determined by just one input cell. In regridding, there are partial overlaps between cells of the input and cells of the output grid; these partial overlaps cause that the value of an output cell can be dependent on multiple input cells. However, it is possible to convert the problem of regridding to a problem of spatial disaggregation. For this purpose, an intermediary grid will be derived. This grid is obtained by considering all the intersections of all cells of the input with all cells of the output. This yields an irregular grid as shown on Fig. 3.2: a grid A and a grid B are combined to form this intermediary grid, the segment grid, S.

The irregular grid S constructed by this intersection has two interesting properties: it partitions all cells of the input grid, and it partitions all cells of the output grid. Consequently, regridding of grid A onto grid B can be done by first performing a spatial disaggregation of grid A into the segment grid S, and then recombining the cells of S to form the cells of the output grid B.

Fig. 3.2 Construction of the segment grid S from input grid A and output grid B

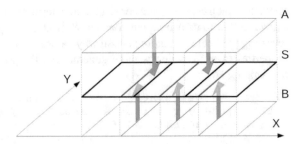

3.1.3 Other Related Applications

3.1.3.1 Data Fusion

The problem of data fusion was explained in Sect. 2.3.4. In data fusion, a single dataset that is an aggregated version of multiple datasets needs to be found. This *aggregation* should be considered in a broad sense: the resulting set should combine the different available datasets into one set that is better, i.e. not only more accurate in the associated values but also in the spatial dimension. The main difference with the spatial disaggregation and the regridding is that there not a single input dataset and proxy grids, but multiple input datasets with which the resulting grid should be consistent. This most likely will change the calculations of the predicates of the rules, but also changes the constraints: the values of the end-result are not necessarily constrained by one of the input grids. While datafusion is not the primary objective at this point, this application is high on the list of future work due to its wide applicability.

3.1.3.2 Identifying Locations

The development of the system for regridding led to another possible application. Identifying locations that meet certain spatial criteria is a complex problem in multi criteria decision making. The concept is very similar to the concept of regridding: using a training set with proxy data and locations that are similar to the ones that need to be identified, a rulebase will be constructed. This rulebase is then later applied to find locations that exhibit a similar connection to the proxy data as those locations in the training set.

3.2 The Rulebase System

The concept of the solution method is to simulate the intelligent reasoning as explained in the previous section, using a fuzzy rulebase system. The details of a fuzzy rulebase system are explained in Chap. 4 in Part II; the concept however is quite intuitive and is introduced here in order to elaborate on the solution method pursued in this work. A fuzzy rulebase basically is a set of rules of the form

```
IF  x_i  IS  L_i^k  AND  ...  x_n  IS  L_n^{k_n}
    THEN  y  IS  L_y^k
```

Here, x_i is an input variable, L_i^k a representation for the linguistic term and y the output variable. In Chap. 4, the mathematics behind the representation of the linguistic terms is further explained, but for now this suffices to consider the concept.

The construction of a rulebase can be done by an expert or by means of a training dataset; this is a dataset that contains datapairs which have values for all variables, including the output variable. In the spatial context, there often is data that can serve as training data. The algorithm presented by the authors in [45] is considered here and is also explained in more detail in Part II, Chap. 6. In the training set, we denote the datapairs $(x_1, x_i, ...x_n; y)$. In addition to these datapairs, every variable needs to have a domain that defines the most possible values for the variable. Over this domain, a number of linguistic terms such as *low* and *high* are defined; here we denote a linguistic term for variable x_i as L_i^k. Each datapair $(x_1, x_i, ...x_n; y)$ is processed and will give rise to a rule that combines the variables x_i with the best matching linguistic term in the antecedent and link y with the best matching linguistic term in the consequent.

```
IF  x_1  IS  L_1^{k_1}  AND  x_i  IS  L_i^{k_i}  AND  ...  x_n  IS  L_n^{k_n}
    THEN  y  IS  L_y^k
```

All these rules are assigned a weight, based on how well the different x_i match and the rules with the best weights are retained. The details on how the data is used to construct the fuzzy rulebase are explained in Chap. 6 in Part II.

Once a rulebase is defined, it can be applied to find an output value y for a given set of values $(x_1, x_i, ...x_n)$. In a fuzzy rulebase system, every rule will be evaluated, and will result in a fuzzy set: the set that matches the linguistic term for y in the consequent of the rule is assigned to y, but weighted with the weight of the rule. The outcomes of all the rules are aggregated into a single fuzzy set for y, which subsequently is defuzzified: a crisp value is extracted from the fuzzy set. Details of this entire procedure are described in Chap. 4 in Part II.

The rulebase approach is chosen as it closely resembles the reasoning of an expert as described earlier in Sect. 3.1. Key to making this approach work is determining appropriate variables, linguistic terms and rule applications. In the spatial context, the output variable y will be chosen so that it relates to the desired value of a cell in the output grid. This means that the rulebase will be applied as many times as there are cells in the output grid, each application of the rulebase determining the value of one cell in the output grid. The input variables used will be defined in such a way that they represent information that relates to the proxy data. These decisions, and related necessary developments in different fields are elaborated on in Part II.

3.3   Novel Developments and Advancements
to the State of the Art

3.3.1   Spatial Processing

The presented method was developed specifically to solve the common spatial problems that occur during the map overlay operation. As such it pushes the state-of-the
art in this field, providing better algorithms to perform spatial disaggregation, regridding and spatial decision making. The novel aspect in performing a disaggregation or
regridding lies in the fact that the developed method allow for additional proxy data,
which can be either gridded or feature based. The system identifies automatically
how the proxy data relates to the given datasets and simulates a reasoning using this
knowledge. In addition, the system allows for expert knowledge to be included.

Apart from the algorithm for spatial disaggregation and regridding, a novel method
for comparing gridded datasets was developed. This new method considers the spatial
similarity between different grids, unlike current methods that mainly consider a
comparison of the numerical values at overlapping locations. By comparing the
spatial similarity, the efficiency of methods that aim to maintain the spatial similarity
can be better determined.

3.3.2   Inference Systems

The methodology makes use of a fuzzy inference system; in order to apply it in a
spatial context, different modifications and extensions needed to be developed. These
are quite general and more broadly applicable and as such, the presented work also
advances the applicability of fuzzy inference systems. The first novel development
is in the construction of Mamdani rulebase systems. Traditionally, the domain of the
variables is determined beforehand and shared over all datapairs; by allowing for
a more accurate definition of the most possible range for the variables, and varying the ranges for different datapairs, a more concise and suitable rulebase can be
constructed. This not only impacts the construction of the rulebase, but also the application of the rulebase. This is a general modification to the construction of rulebase
systems, which can be broadly applicable. The second innovation developed in the
context of the rulebase systems the a new defuzzification method. The new method,
which was necessary for properly interpreting the result in the spatial application,
allows for the defuzzification of multiple fuzzy sets when there is a shared constraint
on the defuzzified value. In our application, the sum of the defuzzified values had
to be a known value, however the approach can be modified for other constraints.
The defuzzification method maximises the lowest occurring membership grade, thus
trying to find the best possible solution given the shape of the fuzzy sets. This is also
more broadly applicable.

Part II
Translating the Spatial Problem

This part explains the development of the artificial intelligent approach. It starts from the concept introduced in the previous part. In order to develop this concept for the spatial problems considered in this book (regridding, spatial disaggregation, identification of locations), it is necessary to translate the spatial problem into a problem that can be solved using a fuzzy rulebase systems. This means translating the problem to a problem consisting of parameters, variables, linguistic terms, domains and rules. The first chapter elaborates on the necessary steps to perform this translation. It starts with the necessary background in fuzzy rulebase systems and defines the notations and nomenclature used throughout. The subsequent chapters deal with different fields that needed to be developed further to translate the spatial problem to a fuzzy rulebase system and apply it in a spatial context. The issues associated with this translation are highlighted, and each of the separate aspects (from the definition of the parameters, over the problems due to locality of the data to specific issues with the construction of the rulebase and with defuzzification) are presented and newly developed solutions proposed in order to develop the algorithm.

Chapter 4
Fuzzy Rulebase Systems

Abstract In this chapter, the fuzzy rulebase systems will be explained. This ranges from how they are constructed from training data, to how they are applied and how the results are interpreted. As the rulebase systems are an application of fuzzy set theory and linguistic terms, the chapter starts with a short introduction on this topic.

4.1 Fuzzy Set Theory

4.1.1 Definition of a Fuzzy Set

Fuzzy rulebase system make use of fuzzy set theory, introduced by Zadeh in [49] as an extension of classic set theory. Where elements in traditional set theory either belong to a set or not belong to a set, fuzzy set theory generalizes this by assigning the elements a membership grade in the range [0, 1] to express the relation of the element with the set.

Definition 4.1 (*fuzzy set*) Consider the universe U. The fuzzy set $\tilde{A}$ has a membership function $\mu_{\tilde{A}}$, defined by

$$\mu_{\tilde{A}} : U \to [0, 1]$$
$$x \mapsto \mu_{\tilde{A}}(x)$$

An example of a fuzzy set over a universe U is shown on Fig. 4.1.

The membership grade can have one of three interpretations: veristic, possibilistic or as degrees of truth; the authors in [7] showed that any other interpretation is equivalent to one of these three. In a veristic interpretation, the membership grades represent the extent to which the elements belong to the set, expressing a partial membership. A real world analogy would be considering the languages a person speaks, where the membership grade serves as indication to how well the language is spoken; a more relevant is example from geographic systems is for example the vegetation in a region, represented by the fuzzy set {(grass, 1), (bushes, 0.8), (trees, 0.2)} to indicate that there is a lot of grass, less bushes, and only few trees. A typical possibilistic

Fig. 4.1 Example of the
fuzzy set A over the
universe U

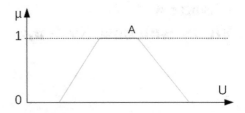

example would be the representation of the age of a person through a fuzzy set when
this age is not exactly known: the person has just one age, and the set expresses which
ages are possible.

Traditional set operations have been extended using specific operations on the
fuzzy sets involved: t-norms act as the union whereas t-conorms (also called s-norms)
act as intersection. T-norms are any function that satisfies the criteria

$$
\begin{aligned}
T(x, y) &= T(y, x) & &\text{(commutativity)}\\
T(a, b) &\le T(c, d) \text{ if } a \le c \text{ and } b \le d & &\text{(monotonicity)}\\
T(a, T(b, c)) &= T(T(a, b), c) & &\text{(associativity)}\\
T(a, 1) &= a & &\text{(identity element)}
\end{aligned}
\tag{4.1}
$$

whereas t-conorms satisfy the properties

$$
\begin{aligned}
S(x, y) &= S(y, x) & &\text{(commutativity)}\\
S(a, b) &\le S(c, d) \text{ if } a \le c \text{ and } b \le d & &\text{(monotonicity)}\\
S(a, S(b, c)) &= S(S(a, b), c) & &\text{(associativity)}\\
S(a, 0) &= a & &\text{(identity element)}
\end{aligned}
\tag{4.2}
$$

T-conorms are in a sense dual to t-norms; for any t-norm, its complementary conorm is
defined by $S(a, b) = 1 - T(1 - a, 1 - b)$. Commonly specific functions are used for
t-norms and t-conorms: Zadeh introduced the minimum as the t-norm ($min(a, b)$) and
the maximum as t-conorm ($max(a, b)$), Lukasiewicz introduced $max(a + b - 1, 0)$
as T-norm and $min(a + b, 1)$ as S-norm. Others are product ($a \times b$) and limited sum
($a + b - a \times b$) as t-norm and t-conorm respectively, and lastly the drastic t-norm
(yields a if $b = 1$, b if $a = 1$ and 0 otherwise) and drastic t-conorm (yields a if $b = 0$,
b if $a = 0$ and 1 otherwise). The algebras obtained by considering certain t-norms
and t-conorms combinations differ in other properties, for which we refer to [8, 23].

4.1.2   α-Cut

Determination of the α-cut of a fuzzy set is an operation commonly used: it removes
all aspects of fuzziness and reverts its (fuzzy) argument to a crisp set. The α-cut of
a fuzzy set is a crisp set that contains all the elements of the fuzzy set for which a

Fig. 4.2 Example of the
α-cut of a fuzzy set A
defined on domain D

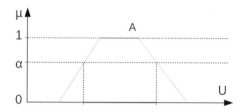

constraint it satisfied: a *strong α-cut* contains the elements with membership grades strictly greater than a given α; a *weak α-cut* contains the elements with membership grades greater than or equal to a given α. This is illustrated on Fig. 4.2.

The *strong α-cut* is defined as:

$$\tilde{A}_{\overline{\alpha}} = \{x \mid \mu_{\tilde{A}}(x) > \alpha\} \tag{4.3}$$

A special case of a strong α-cut is the *support*; this is the strong alpha-cut with threshold 0. This is an important α-cut, as it results all the elements that belong to some extent to the fuzzy set.

$$\tilde{A}_{\overline{0}} = \{x \mid \mu_{\tilde{A}}(x) > 0\} \tag{4.4}$$

The *weak α-cut* is defined as:

$$\tilde{A}_{\alpha} = \{(x, 1) \mid \mu_{\tilde{A}}(x) \geq \alpha\} \tag{4.5}$$

Similarly to the strong α-cut, the weak α-cut has a special case, now for a threshold 1. This α-cut is called the *core*, and returns all the elements that fully belong (membership grade 1) to the given fuzzy set.

$$\tilde{A}_1 = \{x \mid \mu_{\tilde{A}}(x) = 1\} \tag{4.6}$$

4.1.3 Height

The *height* of a fuzzy set returns the highest membership grade that occurs in a fuzzy set. For normed fuzzy sets, this will always equal 1(by definition), but for non-normed sets this can take any value in the range [0, 1]. Formally, the height of a fuzzy set $\tilde{A}$ is defined [8] as:

$$height(\tilde{A}) = \sup_{x}(\mu_{\tilde{A}}(x)) \tag{4.7}$$

Fig. 4.3 Example of the
linguistic terms for low,
medium and high

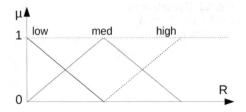

4.2 Fuzzy Numbers and Linguistic Terms

A particular application of fuzzy sets are fuzzy numbers, there are basically fuzzy sets
defined over the domain of real numbers. The fuzzy sets carry a veristic interpretation,
to indicate e.g. all the numbers that satisfy the predicate; for example the fuzzy set
that matches *approximately 5.0* will be such that the closer a number is to 5.0,
the closer its membership grade will be to 1. Operations are possible using Zadeh's
extension principle, and a fuzzy arithmetic has been developed. Several algorithms for
defuzzification, the process of extracting a single crisp value that can be considered
a simplification of the fuzzy set, exist. The most common examples are *center of
gravity* and *mean of max*; defuzzification is considered in more detail in Chap. 7.

Related to this and of interest for the fuzzy rulebase system is the concept of
linguistic terms, allowing for the expression of concepts such al *low* or *high*. The
linguistic term is represented by a fuzzy set as illustrated on Fig. 4.3. The term *low*
expresses all numbers that are considered low; this term has a veristic interpretation
and on the example, 0 is considered to be the lowest number, and greater numbers
are considered to be low to a decreasing extent, all the way to 50; numbers greater
than 50 are no longer considered to be low. The membership functions for medium
and high are interpreted similarly. Any number can be evaluated and can match with
one or more linguistic terms. The number 10 is low to an extent 0.8; the number
40 is low to an extent of 0.20 and medium to an extent 0.80. It is possible to define
modifiers to express concepts such as *very*, but these are not considered further on
in this book. We will use multiple linguistic terms defined over the domain axis to
evaluate values.

4.3 Mamdani Rulebase Systems

A fuzzy inference system, or fuzzy rulebase system is a system that uses fuzzy sets
to represent data and evaluates predicates using simple rules and fuzzy matching
[26]. A lot of research has been done in the field of both Mamdani inference systems
[24] and Sugeno inference systems [37]. In general, fuzzy inference systems employ
fuzzy logic to define a set of rules of the form (in case of Mamdani inference systems)

```
IF  x_i  IS  L_i^k  AND  ...  x_n  IS  L_n^{k_n}
    THEN  y  IS  L_y^k
```

Here, i enumerates the antecedents, x_i is a value for the antecedent i evaluated against linguistic term L_i^k, which is one of the k linguistic terms defined on the universe $[x_i^-, x_i^+]$ of most possible values for x_i. The operator IS in the antecedent evaluates how well the value matches the fuzzy set associated with the linguistic term. The consequent contains the output value (y) and a linguistic term, also represented by a fuzzy set that is suitably defined on the domain of y. The IS operator in the consequent is an assignment, assigning the fuzzy set associated with the linguistic term to y. This assignment is coupled with the lowest membership grade that occurs in the evaluations of the antecedents. Typically, the first step is to determine which linguistic terms are appropriate for the variables, which is done by finding the domain and defining linguistic terms over this domain that provide a meaningful partitioning of this domain.

The input variable is commonly a real number, whereas the linguistic term is represented by means of a fuzzy set defined over a suitable domain of possible values for x_i. In general there are multiple input variables, combined using logical operators *and*. To evaluate a set of input variables, the variable are matched against the linguistic terms; each of these evaluations result in a value in the range [0, 1] that indicates how well the value matches the term. Based on this, y is assigned a linguistic term (and thus a fuzzy set) in its domain. There are multiple rules in the rulebase, and a set of input variables can match multiple rules at the same time, resulting in multiple fuzzy sets for y. All these results for y are aggregated to a single fuzzy set, which is then subsequently defuzzified to yield the crisp result. This is considered the output of the rulebase system. Different algorithms exist to construct fuzzy rulebase systems; they can be supplied by an expert or generated from training data.

The subsequent chapters explain the different steps of rulebase construction, application and interpretation in more detail, along with the developments related to applying the Mamdani fuzzy rulebase systems in a spatial context. This starts with appropriate definitions of the parameters and ranges (Chap. 5) for spatial data. In the spatial context, the fuzzy rulebase system is evaluated for every cell in the output grid. Spatial data can vary in different location, and the standard rulebase construction method cannot take this into account. In Chap. 6, the standard approach for constructing a fuzzy rulebase system from training data is presented, along with the issues that occur with spatial data. The chapter continues with the newly developed method for constructing rulebases that can handle a bigger variation that occurs localized in the data; inline with the results of Chap. 5. The evaluation of a rulebase is performed by evaluating all the rules, which results in a number of fuzzy sets for the output variable. These fuzzy sets are aggregated into one fuzzy set that is the outcome of the rulebase system. For most applications, it is however necessary to have a single, crisp result rather than a fuzzy set. A single value is extracted from the fuzzy set in the defuzzification process, which is explained in detail in Chap. 7. In addition, the fact that the fuzzy rulebase is evaluated for each cell in the output grid creates additional constraints that can be used to improve the defuzzification process and the newly developed methods for defuzzification are also detailed there.

Chapter 5
Parameters and Most Possible Ranges

Abstract The first step in applying a fuzzy rulebase system in the context of spatial data is determining the definition of the variables and linguistic terms In order to use a fuzzy rulebase system for processing spatial data, a number of modifications were necessary. In this chapter, definitions of the parameters that are calculated to determine the variables in the rules are introduced. In addition, appropriate definitions for the possible ranges are also introduced, along with new concepts for most possible ranges that vary with each data pair. This allows for a more optimal posssible range but it will require a modification to the algorithm for construction a rulebase from examples which will be explained in the next chapter. It will also impact the application of the rulebase, where the domains are also not a priori known.

5.1 Parameters

5.1.1 Concept

The concept of the rulebase system-approach is presented in Chap. 3. Translating the spatial problem to a rulebase problem is key to making this concept work. The knowledge available to perform a spatial disaggregation or regridding are the grid that has to be remapped or disaggregated and the target onto which it has to be mapped. In addition, there are one or more spatial datasets that serve as proxy data; these datasets can be either grid based or feature based (Sect. 1.2). The information that can be extract from these datasets needs to be translated into the form of a set of rules:

$$\text{IF} \quad x_i \quad \text{IS} \quad L_i^k \quad \text{AND} \quad \ldots \quad x_n \quad \text{IS} \quad L_n^{k_n}$$
$$\text{THEN} \quad y \quad \text{IS} \quad L_y^k$$

In general, the rule links the variables x_i and their relation to the associated linguistic terms with the linguistic term that is most appropriate for the output term. In the spatial applications, the aim is to have the antecedents link the data to the output. For this purpose, a property derived from both the value and the topological relation of the geometries of the proxy data in relation to the output data is designed.

J. Verstraete, *Artificial Intelligent Methods for Handling Spatial Data*, Studies in Fuzziness and Soft Computing 370, https://doi.org/10.1007/978-3-030-00238-1_5

Similarly, the linguistic term in the rule needs to be defined on a domain of possible values for this derived property. This range, denoted $[x_i^l, x_i^h]$ for a value x_i also needs to be determined; this means that for every variable, three values need to be determined. This triplet has to satisfy a number of criteria:

$$\forall i : x_i^l \leq x_i \leq x_i^h \tag{5.1}$$
$$\nexists x_i \exists c \,|\, x_i / (x_i^h - x_i^l) = c \tag{5.2}$$

The first criteria states that the value of the variable for each output cell is within its associated range. This is not strictly necessary, and some exceptions could be accepted, but such occurrences should be kept to a minimum. The second criteria states that not all evaluations should result in the same outcome: if the value is always at the same relative position in the associated interval, then all data pairs will have the same linguistic term for this variable, which means this term would never contribute.

5.1.2 Defining the Variables

A variable is a value calculated for a given output cell. Different values can be considered; the available information are the different proxy grids and of course the information that can be derived from the geometries. For this purpose, topological operations such as for example intersection, surface area, distance, overlap can be used. While there can be a known connection (e.g. a high weighted overlap with the proxy grid implies high output values), the algorithm is given more options in order to determine different and less trivial connections between proxy data and output data. It should be added that the connection found by the system is a form of correlation, but this does not imply any causality. To calculate a value for a given output cell, first a set of relating cells of the proxy grid are determined. These relating cells can be determined using different spatial queries, for example:

- cells that overlap with the given output cell,
- cells that are within a certain distance of the given output cell,
- cells that overlap with the input cells that overlap with the given output cell,
- all cells of the proxy dataset.

Once a set of relating cells is determined, a computation that involves their associated value will yield the value of the parameter. This computation can be any aggregation, for example:

- the sum,
- the minimum,
- the maximum,
- a weighted sum using amount of overlap as weights,
- a weighted sum using a distance-measure as weights.

Along with a value, it is necessary to determine the range of most possible values. The calculation of this most possible range is considered in the next section.

5.2 Most Possible Ranges

In order to define the linguistic terms that will be used for a given variable, it is necessary to first define an interval that limits the possible values for the variable. In the standard Mamdani rulebase systems ([45] and Sect. 4.3), the most possible range of a variable is shared between the data pairs. However, for the spatial application of the rulebase, this idea was challenged and extended. As such, we distinguish three different approaches to calculate the most possible range for a variable. The first, presented in Sect. 5.2.1 is the approach as traditionally considered in literature. It will be called a *global range*, as the range for a variable is global over all data pairs considered. A variation on this is the *local range*, as presented in Sect. 5.2.2. In this case, ranges for a variable are shared between groups of data pairs, usually data pairs that relate to data that is within a spatial neighbourhood. A different approach, *estimated ranges*, resorts to calculating the range for a variable for each data pair individually, as explained in Sect. 5.2.3. These three categories of range definitions are explained and the resulting rulebases are compared using small examples.

5.2.1 Using a Global Range

5.2.1.1 Definition of the Range

In [45], the first step in constructing the rulebase system is to determine the domain of each of the variables in the rulebase. Determining a good definition of this domain for a variable can be done by first considering all the possible values for the variable in all of the data pairs. The minimum of these values is a suitable value for the lower limit of the most possible range, while the maximum serves as a suitable upper limit. A range defined using these limits satisfies the criteria put forward in (5.2), for all variables and all datapairs. It is considered to be *global* as each data pair uses the same range for the same variable. The next step in the Mamdani rulebase construction is to define the linguistic terms for each variable on their respective range, for example using a standard partitioning as shown on Fig. 5.1.

The downside of global parameters in a spatial context will be explained by means of an example. For this, a small example of spatial disaggregation of a grid A into a grid B will be used. The cells of a grid X will be denoted x_i, with associated values $f(x_i)$. Proxy data C is used to help provide information on the underlying distribution; C is assumed to also be a gridded dataset, but the approach can be modified to work with a feature-based dataset as proxy data. The rulebase approach

Fig. 5.1 Standard
partitioning of the linguistic
terms on a given range: all
linguistic terms (except for
very low and *very high*) are
triangular fuzzy sets

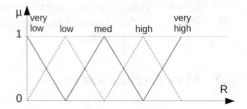

determines these values $f(b_i)$ by considering proxy data, translated as antecedents,
as described in Chap. 3.

```
IF  (property of C relating to b_i) AND ... IS high
      THEN  f(b_i) IS high
IF  (property of C relating to b_i) AND ... IS low
      THEN  f(b_i) IS low
```

This rulebase will be executed for every output cell b_i, resulting in many values
$f(b_i)$ that will be aggregated later. The proxy data C does not have to match grid A
or grid B, nor does it have to partition of either of those—it could even be a feature
based dataset. To define a property that relates C to the underlying distribution of
A, it is possible to use the definition of the geometries of the cells. If C for example
holds similar data as A, but e.g. from a different source or so, the amount of overlap
of the cells of C, combined with the value of those cells, can serve as an indication
for the distribution of the data; this will be the property used here, the value-weighted
amount of overlap denoted $g(b_i)$:

$$x^i = g(b_i) = \sum_{c_j \in C} \frac{S(c_j \cap b_i)}{S(c_j)} \times f(c_j) \qquad (5.3)$$

where $S()$ is the notation of the surface area, $c_j \cap b_i$ is the notation for the geometry
created by the intersection of the geometries of c_j and b_i and $f(c_j)$ is the notation
of the value associated with c_i.

The construction of a rulebase from examples (Chap. 4), and starts from a partitioning of the input spaces and output space. Typically, and also in our example,
the value-weighted overlap of cells of C with the target cell b_i, it is therefore first
necessary to find the domain of most possible values. Using a global range, this is
done by calculating the value for all datapairs in the example, and use the minimum
and maximum to define the range, Fig. 5.2 illustrates which cells are used to calculate the value of the hashed cell. The training set will be denoted A', B' and C' with
elements respectively a'_j, b'_i and c'_k.

Fig. 5.2 The shaded area indicates all the values that will be considered for determining the minimum and maximum possible range for the hashed grid cell

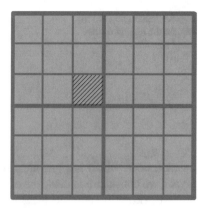

Fig. 5.3 Example of a natural partitioning to define the fuzzy sets for the linguistic terms for low, medium and high in the range $[x^-, x^+]$

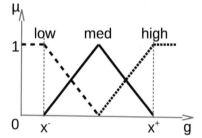

$$x^- = \min_{b'_i \in B'} g(b'_i)$$

$$x^+ = \max_{b'_i \in B'} g(b'_i)$$

The linguistic terms high, low and medium will be defined over the interval $[x^-, x^+]$ as shown on Fig. 5.3. Every datapair (x_i, y_i) in the training set results in a rule:

```
IF  x  IS  L_j^x  THEN  y  IS  L_k^y
```

As before, x and y are the names of the variables for which x_i and y_i are the values and L_j^x, respectively L_k^y in the best matching linguistic term defined on the domain of the variable.

5.2.1.2 Spatial Aspects and Impact

Consider the example as shown on Fig. 5.4. The Figure shows a grid A, consisting of four gridcells a_i, $i = 1..4$ and a grid B with eight gridcells b_{i_j}, $i = 1..4$, $j = 1..2$. The cells b_{i_j} partition the cells a_i. There is a shaded area, more or less located at the

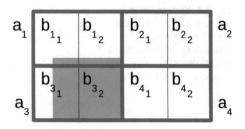

position of a_3 which is a graphical representation of proxy data C. With this grid
layout, we consider two datapairs of the form (x_i, y_i).

Furthermore, we consider the amount of overlap with the shaded area to be indicative for high values in the original grid A. The value of the shaded area can be assumed
to be 1, so the weighted overlap reverts to the amount of overlap. The shaded area
therefor is a graphical representation of the value of the proxy data contained in C.
First, the datapairs associated with b_{1_1} and b_{1_2} are considered. These datapairs have
values for $x_{b_{i_1}}$ and $x_{b_{i_2}}$ (the output values are not illustrated on the figure):

$$(x_{b_{1_1}}, y_{b_{1_1}}) = (0.1, 0.2)$$
$$(x_{b_{1_2}}, y_{b_{1_2}}) = (0.2, 0.8)$$

The cell b_{1_1} overlaps slightly with C, as indicated by the shaded area on Fig. 5.4,
resulting for example in $x_{b_{1_1}} = 0.1$ and $x_{b_{1_2}} = 0.2$. Both of these evaluate to *low*. At
the same time, the output values are $y_{b_{1_1}} = 0.2$ and $y_{b_{1_2}} = 0.8$, which evaluate to *low*
and *high*. Following the algorithm, these two datapairs would yield the rules:

```
IF x IS LOW y IS LOW
IF x IS LOW y IS HIGH
```

The differences between the evaluation of x will cause the rules to have different
weights, and this in favour of the second rule, which is the more consistent one
considering the assumed connection between grids C and A.

However, the disaggregation of a_4 will result in

```
IF x IS HIGH y IS LOW
IF x IS HIGH y IS HIGH
```

with the second rule carrying the highest weight.

In a bigger example, there will also be cases that will add the rules

```
IF x IS LOW y IS LOW
IF x IS HIGH y IS HIGH
```

Combining all these rules will result in a rulebase that has four rules; but they
consider all possible combinations, resulting in outcomes that have a high uncertainty.

The core of the problem lies in the fact that the range of most possible value was chosen too wide. For the datapair in the first example, it is impossible that either $x_{b_{1_1}}$ or $x_{b_{1_2}}$ are greater than 0.3, as the total weighted overlap of a_1 with C is 0.3. In this particular case, a low value (that of a_1) has to be redistributed over both cells b_{1_1} and b_{1_2}. While one of these cells has a bigger portion of the value of a_1, its absolute value is still low. The biggest possible value when looking at the data is 1, which occurs in cell b_{3_2}. The choice of the range [0, 1] is however not suitable for this datapair and causes the construction of rules that are less effective at solving the problem.

This is the easiest way of defining a value and associated range, but due to the fact that they cannot account for local differences, the performance tends to be limited. Performance-wise, the calculation of the parameters can be sped up by caching the range, as it is the same for all datasets.

5.2.2 Using a Local Range

In order to accommodate for local differences, a different idea was put forward: rather than considering all the datapairs for determining the range of possible values, only those datapairs that relate to cells that are in the neighbourhood of the current cell are considered. A range defined like this is called *local* as the relating cells of the proxy grid are determined from values of cells in the local neighbourhood. Determining the range is similar to a global range, but restricted to the region of the map in the area of the output cell. This can be achieved by defining a buffer around the output cell and considering all the cells that overlap with this buffer, or by using other grids to help define the neighbourhood. The neighbourhood can for example be defined by all the output cells that overlap with the input cell a_j that overlaps the considered output cell. Both these options are illustrated on Fig. 5.5. The latter option would employ the formulas below to calculate the most possible range for a given cell.

$$x_{b_i}^- = \min_{b_j' \in B' | b_j' \cap a_k' \ in A' \neq \emptyset} g(b_i') \quad \text{WHERE} \quad b_i' \cap a_k' \neq \emptyset$$

$$x_{b_i}^+ = \max_{b_j' \in B' | b_j' \cap a_k' \ in A' \neq \emptyset} g(b_i') \quad \text{WHERE} \quad b_i' \cap a_k' \neq \emptyset$$

As a result, a given datapair (x^{b_i}, y^{b_i}) will have its own range for x^{b_i}: $[x_{b_i}^-, x_{b_i}^+]$. The range for the output variable y^{b_i} is not considered here in further detail, but can be calculated in a similar way.

In order to apply the rulebase with the algorithm in [45], it is necessary for all datapairs to have the same most possible range. By rescaling every value of x^{b_i} using its individual range $[x_{b_i}^-, x_{b_i}^+]$ this common range is achieved; arbitrarily, the range [0, 1] was chosen. The scaled function will be used determine values as input for the rulebase and is denoted as:

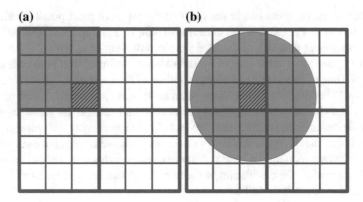

Fig. 5.5 Two examples to determine the cells considered for determining the minimum and maximum possible range for the hashed cell, indicated by the shaded area: **a** using those cells that overlap the same input, **b** using the cells that overlap a given neighbourhood

$$g'(b_i) = \frac{x^{b_i} - x_{b_i}^-}{x_{b_i}^+ - x_{b_i}^-} \tag{5.4}$$

Using this definition, the rulebase construction method can be applied as in the previous section. The same example on Fig. 5.4, the two datapairs $(x_{b_{1_1}}, y_{b_{1_1}}) = (0.1, 0.2)$ and $(x_{b_{1_2}}, y_{b_{1_2}}) = (0.2, 0.8)$ will now first have to be rescaled. This yields:

$$g'(b_{i_1}) = \frac{x^{b_{i_1}} - x_{b_{i_1}}^-}{x_{b_{i_1}}^+ - x_{b_{i_1}}^-} = \frac{0.1 - 0.1}{0.2 - 0.1} = 0$$

$$g'(b_{i_2}) = \frac{x^{b_{i_2}} - x_{b_{i_2}}^-}{x_{b_{i_2}}^+ - x_{b_{i_2}}^-} = \frac{0.2 - 0.1}{0.2 - 0.1} = 1$$

In an example where there would be more than two cells partitioning the input cell a_j, other cells can have values that differ from 0 or 1. This transformation of the input variable results in two new datapairs that will be used for the rulebase construction: $(0, 0.2)$ and $(1, 0.8)$. These datapairs will evaluate respectively to (*low*, *low*) and (*high*, *high*), yielding the rules:

```
IF x IS LOW y IS LOW
IF x IS HIGH y IS HIGH
```

While it still is possible for different rules to be constructed, such rules will stem from datapairs that are inconsistent with the general assumption on the relation between A and C. In Chap. 6, an algorithm to construct and apply the rulebase without the need for rescaling is presented.

5.2.3 Estimated Ranges

The most possible range is a property which is connected to the output cell for which the value is calculated, so it is possible to consider calculating a most possible range rather than deriving it from the parameter values of neighbouring cells.

Consider the output grid B and a proxy grid C. The spatial distribution of gridded data is not known, but this is also the case for the data in grid C. The different options for the spatial distribution of the data in this grid can therefor be used to determine the limits. The value-weighted overlap, which is used as the value, matches the assumption that data in the gridcells of C is uniformly distributed. However, a distribution as shown on Fig. 5.6a can be considered: the underlying spatial distribution of grid C can be such that as much as possible lies outside of the cell b_i that is considered. This is only possible for cells c_k that have a partial overlap with b_i; what remains is the overlap of the cells of C that are fully contained. Such a distribution effectively results in the lowest possible value for the parameter for b_i.

$$
\begin{aligned}
x_{b_i'}^- &= \sum_{c_j \in C | c_j \subset b_i} \frac{S(c_j \cap b_i)}{S(c_j)} \times f(c_j) \\
&= \sum_{c_j \in C | c_j \subset b_i} f(c_j)
\end{aligned}
$$

Similarly, the spatial distribution of the data in the cells of C could be such that it maximizes the amount of data mapped in the considered cell b_i, as shown on Fig. 5.6b. The underlying spatial distribution of the partially overlapping cells is in this case such that the data is fully contained in the considered cell b_j; the maximum therefore is the sum of all overlapping cells.

$$
x_{b_i'}^+ = \sum_{c_j \in C | c_j \cap b_i \neq \emptyset} f(c_j)
$$

This range the benefit that it effectively evaluates the value (in this case obtained through value-weighted overlap) against the realistically possible values, and this is optimized for each datapair. On the other hand, the property has its limits: when there are no partially overlapping cells between b_j and C, the value for the upper limit will be the total sum, which only in specific cases can be reached. Similarly, when the cells of C are very big in size compared to those of B, the range may be skewed too much towards high values, resulting in the evaluation of all values of b_j to be rather low.

Estimated ranges are more difficult to construct than local parameters, and are more prone to unforeseen side-effects. However, a properly defined estimated range can capture an expert's interpretation on how grids relate to one another in much

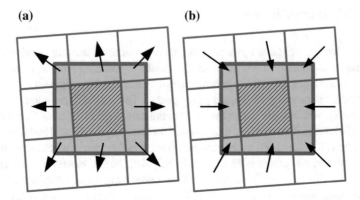

(a)   **(b)**

Fig. 5.6 Interpretation of the estimated minimum and maximum, as respectively contained cells (**a**) and overlapping cells (**b**)

more advanced ways. It is difficult to illustrate the estimated ranges on a simple example as the one on Fig. 5.4; for this small example it would perform similar as to the local range, but this does not illustrate the difference in more general cases.

Using estimated ranges to construct and evaluate a rulebase is similar to using local ranges and requires the rescaling of the value as mentioned on Sect. 5.2.2, or the rulebase construction method from Chap. 6. The assessment whether or not the rules generated from the examples are consistent with the assumed connection with the proxy data is more difficult than when using local ranges.

5.3   Selecting Parameters

5.3.1   *Input Parameters*

There is a the multitude of possibilities to define a parameters due to the different possibilities for deriving a value from proxy data and the different combinations for defining an appropriate range. Consequently, an obvious question arises: what are the best parameters? As there is no set of parameters that is the best for all cases, it is necessary to identify which parameters are the most suitable for the given situation. The value and range of a parameter are such that it should relate to the ideal output. This property can be exploited to judge the quality of a parameter: for a variable v_j, a grid X that has the same layout as the output grid B will be constructed. Every cell c_i corresponds with a datapair $(x_j^{(i)}; y^{(i)})$. The range of $x_j^{(i)}$ is $[x_j^{(i)-}, x_j^{(i)+}]$ The values $f(x_i^j)$ associated with the cells of X_j are defined as

$$f(c_i) = \begin{cases} \frac{v_i - x_j^{(i)-}}{x_j^{(i)-} - x_j^{(i)+}} & \text{if } x_j^{(i)+} \neq x_j^{(i)-} \\ 0 & \text{elsewhere} \end{cases} \qquad (5.5)$$

Subsequently, this grid is scaled so that the total value is the same as that of the output grid. If the parameter relates to the output, then this constructed grid should resemble the output grid. To evaluate this, the developed method for ranking grids (Chap. 8) will be used. The more the grid resembles the output grid, the better the parameter is considered to be. Both a parameter definition and a range definition incorporate one or more proxy layers and/or the input layer or even the geometry of the cells in the output layers.

Algorithm 1 Algorithm from parameter selection

1: grid INPUT
2: grid OUTPUT
3: set of data PROXY
4: **for all** each parameter definition P and range definition R **do**
5:    grid TEST = new grid with same layout as OUTPUT
6:    **for all** cell c in OUTPUT **do**
7:       calculate P(c)   ▷ value
8:       calculate R(c)   ▷ range
9:       rescale P(c) using R(c)
10:       c in TEST = P(c)
11:    **end for**
12:    match TEST with OUTPUT as reference   ▷ indicates quality
13: **end for**

The above algorithm provides a ranking for the different parameters. While it makes sense to select those parameters that rank the highest, the side effect of this may be that some proxy data is ignored completely. In addition, parameters that are defined on the same grids may not be fully independent of one another, so making sure other proxy data is not ignored is important when selecting the parameters.

5.3.2 Output Parameter

When the rulebase is applied, it is also necessary to define a range for the output parameter; and again different choices of ranges are possible. Consequently, an appropriate range also needs to be determined for the output parameter. This is done in the same way as for the input parameters when the rulebase is constructed: at construction time, a possible output parameter, which includes the calculated value and the range is computed. Similar to how input parameters are scaled and tested against the ideal output parameter, this value is scaled within its range, mapped onto the grid and this grid compared with the ideal output grid. This procedure allows to select the best parameter for the output.

Chapter 6
Rulebase Construction

Abstract From the previous chapter, it became clear that global ranges are not always suitable for applying a rulebase in spatial applications and that better ranges can be determined. Here, a modification to allow a fuzzy rulebase system to be used for processing spatial data is presented. This is a modification to the algorithm for construction a rulebase from examples, eliminating the requirement that the domains of the variable need to be known in advance, provided it can be calculated for every variable. This also impacts the application of the rulebase, where the domains are also not a priori known. Unlike with a normal rescaling, the resulting fuzzy set is correctly defined on the domain of the result, which is beneficial for subsequent processing.

6.1 Introduction

As explained in Sect. 4.3, fuzzy rulebase systems allow to model connections between input values and output values without a priori knowledge on how input and output relate. In general, the construction of a fuzzy rulebase system requires partitioning the input and output spaces for the variables x_i and y; these partitions are then used to define the linguistic terms L_i^k for the variables. The spaces are not limiting, in the sense that values are allowed to be outside of them; however all values outside of the spaces are treated the same as the limits of the spaces and thus these values cannot be differentiated further. This limits the expressiveness of the rules. Making these spaces too small causes too many values to fall outside of them; making the space too big lowers the ability to distinguish between different values, lowering the performance of the system. In [45], the authors present an algorithm to generate the rules using an iterative process from data pairs and to later select and combine the rules based on evaluations of antecedents to result in a rulebase of manageable dimensions. In [5], a method using hinging hyperplanes is used to construct the fuzzy rulebase. The authors in [6] discuss and compare different approaches to generate the rules: the first uses a genetic algorithm combined with heuristics to perform the selection of the remaining rules, based on the degree of coverage, while the second method discussed uses a Bayesian classifier to select the rules. Common to all these approaches, is that

© Springer Nature Switzerland AG 2019

J. Verstraete, *Artificial Intelligent Methods for Handling Spatial Data*, Studies in Fuzziness and Soft Computing 370, https://doi.org/10.1007/978-3-030-00238-1_6

little attention is given to the linguistic terms and how their distribution affects the rules: the assumption is made that for every antecedent i, the same universe $[x_i^-, x_i^+]$ of most possible values is used for the definition of the linguistic terms, which will also be the same for every antecedent i. In [29], the authors continue the work of [31] (in Japanese); both methods use genetic algorithms to determine the fuzzy partitioning of the input space. These methods are capable of providing a customized definition of the linguistic terms based on the data in the training set. These methods optimize the number of linguistic terms needed to still have enough possibilities for distinguishing the values of the different data pairs. The partitioning of the input space for each antecedent is still considered globally: every data pair will use the same partitioning for the same variable. In our research in the application of a fuzzy rulebase system for spatial disaggregation ([42]), the universes and values for the same antecedent in the rulebase exhibited huge differences between different data pairs, even though the general format of a rule (*if value is high then...*) would still hold over these data pairs when the optimal universe is considered and the linguistic terms are distributed similarly over the appropriate universes. As was shown in Chap. 5, in this particular situation, a traditional a priori division of the universes for each antecedent is insufficient to distinguish values, decreases the impact of the rule and results in poor evaluation of values when creating or applying the rules.

In the next section, the algorithm presented in [45] is explained, as this serves as the basis for our algorithm. The subsequent section explains the developed algorithm for constructing a fuzzy rulebase using optimized possible ranges.

6.2 Generating Rules from Examples with Constant Spaces

The standard algorithm for the construction of a rulebase is presented in [45]. The steps to this algorithm are summarized below, in order to highlight the differences with the rulebase construction method presented further on. Too construct a rulebase from examples, the set of input-output datapairs denoted:

$$(x_1^{(1)}, x_2^{(1)}; y^{(1)}), (x_1^{(2)}, x_2^{(2)}; y^{(2)}), ... \qquad (6.1)$$

$x_j^{(i)}$ denotes the j-th value of data pair i.

Generating the rulebase is a stepwise process consisting of the following steps shown in Algorithm 2.

The domain intervals for each of the variables are defined as the most possible interval from which the variable can have values. The domain intervals for an input variable x_i will be denoted $[x_i^-, x_i^+]$; the domain for the output variable will be $[y^-, y^+]$.

It is possible for values to be outside of this interval, but in those cases they will be treated the same as the edge of the interval. In step 1 of the process, the linguistic terms

---

Algorithm 4 Algorithm for rulebase generation with constant spaces

---
1: Divide input and output spaces into fuzzy regions
2: Define linguistic terms on these spaces
3: **for all** datapair **do**
4:    Generate fuzzy rule
5:    Assign a degree to the each rule
6: **end for**
7: Create combined fuzzy rulebase
8: Determine a mapping based on the combined fuzzy rulebase

---

for a variable are defined on these intervals, and for every i, the interval $[x_i^-, x_i^+]$ is independent of the data pair and consequently uses the same linguistic terms.

With the intervals and linguistic terms defined, the values of each variable of data pair are evaluated against these terms in step 2. The linguistic term that results in the highest membership grade is retained and associated with the variable to form a rule. Each data pair yields one rule.

Different data pairs can result in the same used linguistic terms for all variables. In order to combine the rules and eliminate conflicting rules, step 3 assigns degrees to different rules. The degree of a rule is defined as the product of the membership grades the different variables. Optionally, this can be multiplied with a degree indicating how good the data pair is (this degree can be assigned by a human expert).

After step 3, there is a set of rules with as many rules as there are data pairs; all rules have a degree and some rules have the same linguistic terms for the same variables. In step 4, only those rules with the highest degrees are retained.

The last step in the rulebase construction is the creation of a mapping function that uses defuzzication of the resulting fuzzy sets using the centroid. This mapping function allows for a very fast application of the rulebase.

6.3 Generating Rules from Examples: Variable Spaces

6.3.1 What Are Variable Spaces?

Consider the situation where, for each variable in each data pair, it is possible to determine an individual, optimal most possible range, for example using a local range or estimated range as described in Chap. 5. For the i-th datapair, this can be denoted:

$$(x_1^{(i)}, x_2^{(i)}; y^{(i)}) \tag{6.2}$$

with specific most possible ranges:

Fig. 6.1 Optimal
distribution of linguistic term
for different ranges

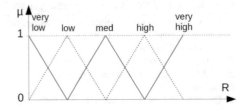

$$x_1^{(i)} : [x_1^{(i)-}, x_1^{(i)+}]$$ (6.3)
$$x_2^{(i)} : [x_2^{(i)-}, x_2^{(i)+}]$$ (6.4)
$$y^{(i)} : [y^{(i)-}, y^{(i)+}]$$ (6.5)

To illustrate the impact of using these more appropriate ranges, the example below
will be used throughout this section. Consider the two data pairs given below

$$(x_1^{(1)}; y^{(1)}) = (10; 30)$$
$$(x_1^{(2)}; y^{(2)}) = (100; 80)$$ (6.6)

with the most possible ranges known to be

$$x_1^{(1)} : [0, 100]$$ (6.7)
$$y^{(1)} : [0, 100]$$ (6.8)
$$x_1^{(2)} : [90, 170]$$ (6.9)
$$y^{(2)} : [70, 120]$$ (6.10)

When the linguistic terms are defined as on Fig. 6.1, then 10 is *very low* (to extent
0.6) and *low* (to extent 0.4) for the first data pair, as the possible ranges here are
[0, 100]; similarly for the second data pair, 100 is *very low* and *low* (both to extent 0.5)
because the possible range is [90, 170] and respectively [70, 120]. Both datapairs,
using their optimal ranges associate the term *very low* with the input variable and
the term *low* with the output variable, thus both leading to the same rule.

6.3.2   Combining the Ranges

Using the standard algorithm described above (Sect. 6.2) implies that it is necessary
to find a single most possible range for each variable, which will then be used for all
datapairs. The logical way to achieve this is by considering the smallest interval that
encompasses all ranges:

Fig. 6.2 Linguistic terms defined for specific most possible ranges for each data pair: **a** shows definitions for antecedent $x_1^{(1)}$ which has range [0, 100]; **b** shows definitions for antecedent $x_1^{(2)}$ which has most possible range [90, 170]

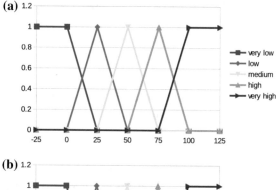

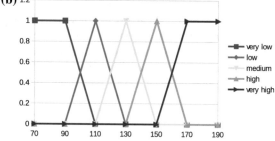

Fig. 6.3 Linguistic terms defined for combined most possible range [0, 170] of the antecedents $x_1^{(1)}$ and $x_1^{(2)}$

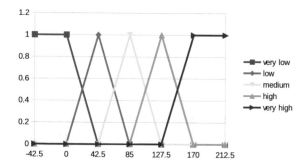

$$[x_j^-, x_j^+] = [\min_i \{x_j^{(i)-}\}, \max_i \{x_j^{(i)+}\}] \tag{6.11}$$

$$[y^-, y^+] = [\min_i \{y^{(i)-}\}, \max_i \{y^{(i)-}\}] \tag{6.12}$$

Using these definitions, the algorithm can be applied immediately. For the example given above, the most possible range for x_1 in all data pairs will be $[x_1^-, x_1^+] = [0, 170]$. Using the five linguistic terms defined on this interval as shown in Fig. 6.3, the value of 10 in the first data pair will evaluate to *very low* and to *low* (respectively to extent 0.76 and 0.24), while the value 100 in the second will evaluate to *medium* (to extent 0.65) and to *high* (to extent 0.35).

The resulting linguistic term differ from those when using the optimal ranges (Sect. 6.3.1), but of course the terms and consequently the rulebase carry a different interpretation. This in turn impacts the rules that will be constructed, and conse-

quently affects the rules that are filtered out in step 4 of the algorithm. The example
in Chap. 5 illustrates better why this rulebase may not be as suitable; the underlying
reason is that does not use optimal information.

6.3.3   Rescaling Most Possible Range Prior to Rulebase Construction

To solve the variability of the ranges of variables between datapairs, a possible
approach is to rescale the values and the ranges, prior to constructing the rulebase.
This will convert the datapairs to a new set of datapairs where variable shares the
possible range. For each input variable j, the rescaled domain for the variables can
be denoted $[x'^{-}_j, x'^{+}_j]$. The domain of the output variable will be denoted $[y'^{-}, y'^{+}]$.
These domains are arbitrarily chosen to [0, 100]; the choice of this has no impact on
the end result as these values are only used for intermediate calculations.

$$[x'^{-}_j, x'^{+}_j] = [0, 100] \tag{6.13}$$

$$[y'^{-}, y'^{+}] = [0, 100] \tag{6.14}$$

The rescaling of the values $x^{(i)}_j$ to $x'^{(i)}_j$ and $y^{(i)}$ to $y'^{(i)}$ is straight forward:

$$
\begin{aligned}
x'^{(i)}_j &= (x^{(i)}_j - x^{(i)-}_j) * \frac{x^{+}_j - x^{-}_j}{x^{(i)+}_j - x^{(i)-}_j} + x^{-}_j \\
y'^{(i)} &= (y^{(i)} - y^{(i)-}) * \frac{y^{+} - y^{-}}{y^{(i)+} - y^{(i)-}} + y^{-}
\end{aligned}
\tag{6.15}
$$

Using this new dataset, the standard algorithm can be applied. Consequently, the
rulebase can now be constructed using the standard approach. For the small example,
this results in the following:

$$
\begin{aligned}
(x'^{(1)}_1; y'^{(1)}) &= (10; 30) \\
(x'^{(2)}_1; y'^{(2)}) &= (12.5; 20)
\end{aligned}
\tag{6.16}
$$

with most possible ranges:

$$x' : [0, 100] \tag{6.17}$$

$$y : [0, 100] \tag{6.18}$$

The evaluation against the linguistic terms (Fig. 6.4) of the rescaled values matches
the evaluation of the original values in the introduction: 10 is *very low* (to extent 0.6)
and *low* (to extent 0.4) for the first data pair, 12.5 is *very low* and *low* (both to extent
0.5) for the second.

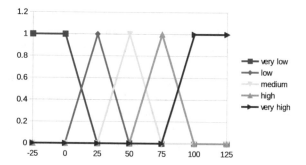

Fig. 6.4 Linguistic terms defined for the common range to which all specific ranges are scaled; the range [0, 100] is arbitrary

The rulebase is defined on scaled values; when applying it, the input data will have to be rescaled, while the resulting output will also have to be rescaled afterwards. For the values $(x_j^{(i)})$, first the ranges $[x_j^{(i)-}, x_j^{(i)+}]$ are determined, after which the values can be rescaled to fit the global ranges (Eq. 6.12); this yields:

$$x'^{(i)}_j = (x_j^{(i)} - x_j^{(i)-}) * \frac{x_j^+ - x_j^-}{x_j^{(i)+} - x_j^{(i)-}} + x_j^- \tag{6.19}$$

The outcome is a fuzzy set defined on the range $[y'^{(i)-}, y'^{(i)+}]$. The defuzzification of this fuzzy set, y' can now be rescaled using the most possible interval $[y^{(i)-}, y^{(i)+}]$ for the output value y the range, using:

$$y^{(i)} = (y'^{(i)} - y^-) * \frac{y^{(i)+} - y^{(i)-}}{y^+ - y^-} + y^{(i)-} \tag{6.20}$$

This application and subsequent rescaling yields a crisp result y which is the evaluation of a given datapair. This approach gives a result that is in line with example from the introduction.

6.4 Developed Algorithm for Learning from Examples with Variable Spaces

The solutions to dealing with datapairs with variable spaces as presented in the previous section exhibit some shortcomings. The first option, considering the union of the most possible ranges for a variable as the most possible range for this variable, changes with which linguistic terms the values match. As shown in the example, it limits the expressibility of the rulebase and not necessarily results in the desired outcome considering the full knowledge of the most suitable ranges.

The second option, rescaling the datapairs when learning and applying the rulebase requires the end result to also be scaled back to match the correct range. In order to scale the datapairs, all ranges have to be known from the start; otherwise there is the risk that some datapairs have ranges that would not be scaled well. In addition, scaling the output requires defuzzifying it, but we are interested in the fuzzy set. To overcome these two drawbacks, a novel algorithm was developed; this will be presented below.

The key to the algorithm is disconnecting the domains from the linguistic terms: each datapair will have its own linguistic terms defined; the only common property between the datapairs is the number of the linguistic terms, and—to keep things simple—the shape of the fuzzy sets associated with them. To achieve this, the initial steps from the rulebase construction in constant spaces are replaced by the following steps:

Algorithm 5 Algorithm for rulebase generation with variable spaces

---
1: **for all** datapair **do**
2:     Determine input and output spaces
3:     Define linguistic terms on these spaces
4:     Generate fuzzy rule
5:     Assign a degree to the each rule
6: **end for**
7: Create combined fuzzy rulebase
8: Determine a mapping based on the combined fuzzy rulebase

For all data pairs, the linguistic terms for a value are defined in a similar way: for each antecedent, the most probable range is determined, and over this range a predefined number of linguistic terms will be defined. For a given data pair $(x_1^{(i)}, x_2^{(i)}; y^{(i)})$, in step 1.c, the algorithm would generate a rule using the most possible ranges

$$x_1^{(i)} : [x_1^{(i)-}, x_1^{(i)+}]$$ (6.21)

$$x_2^{(i)} : [x_2^{(i)-}, x_2^{(i)+}]$$ (6.22)

$$y^{(i)} : [y^{(i)-}, y^{(i)+}]$$ (6.23)

and using linguistic terms defined on these most possible ranges (in step 1.b).

Consider two datapairs $(x_1^{(1)}, x_2^{(1)}; y^{(1)})$ and $(x_1^{(2)}, x_2^{(2)}; y^{(2)})$. For the first data pair, the intervals are: $[x_1^{(1)-}, x_1^{(1)+}]$, $[x_2^{(1)-}, x_2^{(1)+}]$ and $[y^{(1)-}, y^{(1)+}]$; for the second data pair the intervals are $[x_2^{(1)-}, x_2^{(1)+}]$, $[x_2^{(2)-}, x_2^{(2)+}]$ and $[y^{(2)-}, y^{(2)+}]$. The rulebase matches values with a relevant linguistic term, which should only matter for this datapair; for the examples as shown on Fig. 6.2.

To illustrate this, consider the smaller example with one input variable from the introduction:

$$(x_1^{(1)}; y^{(1)}) = (10; 30)$$
$$(x_1^{(2)}; y^{(2)}) = (100; 80)$$ (6.24)

with the most possible ranges as before defined as:

$$x_1^{(1)} : [0, 100] \tag{6.25}$$
$$y^{(1)} : [0, 100] \tag{6.26}$$
$$x_1^{(2)} : [90, 170] \tag{6.27}$$
$$y^{(2)} : [70, 120] \tag{6.28}$$

The linguistic terms are now defined for each data pair individually, as shown on Fig. 6.4; which matches the description in Sect. 6.3.1. As a result, 10 is *very low* (to extent 0.6) and *low* (to extent 0.4) for the first data pair, 100 is *very low* and *low* (both to extent 0.5) for the second. The further processing steps in constructing the rulebases do not require the actual definitions of the linguistic terms, merely the linguistic terms themselves, allowing the rules from all data pairs to be combined and selected as before.

Applying the rulebase implies that for every variable a most possible range is determined. Important here is that also the most possible range for the output variable has to be determined. The input variables are matched against their set of specific linguistic terms, resulting in a linguistic term for the output, for each rule that matches. These linguistic terms are matched with the associated fuzzy sets defined on the calculated most optimal range of the output variable and aggregated. The resulting fuzzy set is a fuzzy set that represents the outcome of the fuzzy rulebase, on the domain suitable for the variable.

Chapter 7
Constrained Defuzzification

Abstract The output of a fuzzy rulebase system is a fuzzy set. In a typical application of a fuzzy rulebase system, this fuzzy set is sufficient for the end result, and
defuzzification of it results in the final outcome. However, in the spatial application,
there are a number of rulebase applications that together form the end result. This
chapter elaborates on the interpretation of the results and the postprocessing necessary to properly interpret the results and explains the shortcomings of the current
methodology, justifying the development of new methods explained in this chapter.
This concerns the defuzzification of fuzzy sets, where a method was developed to
defuzzify multiple fuzzy sets in parallel, under the condition that the result should
sum up to a given value. This constrained defuzzification allows for better results in
the context of the spatial processing, but can also have wider applications.

7.1 Defuzzification

7.1.1 Definition

In fuzzy sets, membership functions represent possible values or approximate values. For many applications, it is necessary to extract a single crisp value from the
fuzzy set [40]. Defuzzification is the process where this single value is determined.
This is an important step to yield a value that is easier to comprehend, or usable in
subsequent calculations. All values of the domain of the fuzzy set are considered and
the membership function is used to help decide which value will be selected. While
it may seem natural to search the core of the fuzzy set (Eq. 4.6) for a suitable value,
even this can be argued against in the case of e.g. very skewed fuzzy sets. It suffices
to compare a left-skewed and a right-skewed triangular fuzzy set that have the same
core: the use of the core would completely dismiss the uncertainty represented by
the membership function.

Commonly used functions for defuzzying fuzzy sets are the Center of Gravity and
the Mean of Max [50]. Center Of Gravity (COG) is defined as:

$$COG(\tilde{A}) = \frac{\sum_{x_{min}}^{x_{max}} x\mu_{\tilde{A}}(x)}{\sum_{x_{min}}^{x_{max}} \mu_{\tilde{A}}(x)} \qquad (7.1)$$

COG considers the value that matches the center of gravity of the fuzzy set as the most representative value. This value not necessarily has the highest occurring membership grade, but does take into account the shape of the entire fuzzy set. Mean Of Max (MeOM) is defined as

$$MeOM(\tilde{A}) = \frac{\sum_{x \in \tilde{A}_\alpha} x}{|\tilde{A}_\alpha)|} \qquad (7.2)$$

where $\alpha = \text{height}(\tilde{A})$ Mean of max considers the mean value of all elements that have the highest membership grade. This also does not guarantee it has the highest occurring membership grade (e.g. if the fuzzy set is non-convex and has two disconnected intervals at its height), and completely ignores the shape of the fuzzy set.

7.1.2   Criteria

Different methods for defuzzification exist; in [40], a number of criteria by which they can be judged, are presented. The authors consider different properties for different types of domains, from arbitrary universes to fuzzy quantities [49]. In our application, only fuzzy quantities (fuzzy sets over the real domain) are considered, and the presented algorithm is also only considered for such fuzzy sets. The authors [40] lists multiple criteria by which defuzzifiers can be evaluated, those of interest for our application are summarized below:

1. Core selection or semantically correct defuzzification: the defuzzified value is a value with highest occurring membership grade.
2. Scale invariance: the scale of the domain does not influence the relative position of the defuzzified value (relative scale concerns translation of the fuzzy set, ratio scale concerns scaling of the unit, interval scale combines both relative and ratio scale).
3. Monotony: the defuzzified value of a fuzzy set that has greater values should be greater.
4. x-Translation: if the fuzzy set is translated, the relative position of the defuzzified value should remain.
5. x-Scaling: if domain values are multiplied with a constant factor, the relative position of the defuzzified value should remain.
6. Continuity or robustness: a small change in membership grades should not yield a big change in defuzzified value.

Center of gravity for example, does not satisfy the core selection criteria (not even for non-convex fuzzy sets), as the method is not guaranteed to return an element with the highest occurring membership grade. It also does not satisfy any aspect of scale invariance. MeanOfMax on the other hand does not satisfy the core selection criteria in the event of non-convex fuzzy sets, but it is guaranteed to satisfy it if the fuzzy set is convex. It also satisfies all other criteria.

The criteria are not mandatory, as the examples show, but knowing which criteria are satisfied gives an insight in the workings of the defuzzifier. The developed algorithm for simultaneous constrained defuzzification will be evaluated against these criteria.

7.2 Developed Simultaneous Constrained Defuzzification

7.2.1 Introduction

Defuzzification is usually an operation performed on a single fuzzy set without outside influences. The authors in [47, 48] proposed the defuzzification of a fuzzy set under specific constraints. These constraints are imposed on the domain of the fuzzy set, limiting the possibilities for defuzzification. While this allows for further adjustment of the behaviour, it still considers defuzzification of a single fuzzy set at a time.

Consider that there is a shared constraint on the defuzzified values: the defuzzified values of the fuzzy sets involved should sum up to a given value. Traditionally, one would defuzzify the fuzzy sets separately, and scale the values. As an example, consider two fuzzy sets A_1 and A_2 as illustrated on Fig. 7.1. Next, suppose that the common constraint is such that the values $MeOM_1 + \Delta_1$ and $MeOM_2 + \Delta_2$ add up to the given constraint. By considering these values, the lowest occurring membership grade is $\min(\mu_1, \mu_2)$. However, if the values $MeOM_1 + \Delta_1'$ and $MeOM_2 + \Delta_2'$ are considered, the constraint still holds, but the lowest occurring membership grade now is $\min(\mu_1', \mu_2')$, which is greater than $\min(\mu_1, \mu_2)$.

Both fuzzy sets, and consequently their defuzzified values are part of the solution. Better solutions have higher membership grades, thus but as both results together are necessary, maximizing the minimum membership grade will result in a better overall solution. This is what is achieved in the second option, resulting in a higher minimum membership grade, indicating that this is a better overall solution to the problem.

The developed method for constraint defuzzification aims at achieving an optimal solution when a shared constraint is imposed on the defuzzified values. In the next section, we consider defuzzifying multiple fuzzy sets under the constraint that the sum of the defuzzified values is known. When the constraints involve multiple fuzzy sets, treating the fuzzy sets individually is no longer an option.

Fig. 7.1 Example to
illustrate the benefits of
constraint disaggregation:
$\Delta_1' + \Delta_2' = \Delta_1 + \Delta_2$, but
$\min(\mu_1', \mu_2') > \min(\mu_1, \mu_2)$

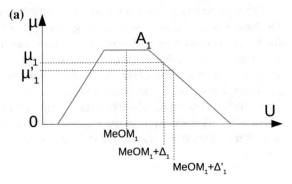

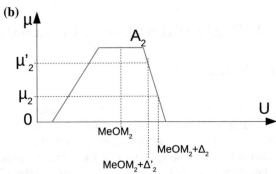

Fig. 7.2 Illustration of the
spatial disaggregation: grid
A needs to be mapped onto
grid B

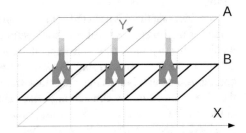

   The constraint defuzzification is not just an exercise without application; it was
developed specifically for the developed spatial disaggregation approach. Particularly
in this context, the constraint defuzzification has benefits. Consider the example of
spatial disaggregation on Fig. 7.2 as first presented in Sect. 2.2.3.

   In the developed approach, the fuzzy rulebase system will be applied for every
output cell; this implies that every output cell will receive a fuzzy set which will have
to be defuzzified. However, the output cells form a partitioning of the input cell, and
as such they should aggregate to value of the input cell. In the considered example
using absolute values, this means:

$$\forall j, \sum_{i|a_j \cap b_i \neq \emptyset} f(b_i) = f(a_j) \tag{7.3}$$

Improving the method for defuzzification under this constraint will improve the overall result, as illustrated with the example on Fig. 7.1.

7.2.2 Definition

In the considered concept for processing gridded spatial data, the defuzzification process could be improved by considering multiple fuzzy sets at the same time: these fuzzy sets are connected by the fact that their defuzzified values should sum up to a given value. The concept of the simultaneous defuzzification is that the best crisp values are such that the lowest membership grade of these values in their respective fuzzy sets is maximized. These two conditions translate to:

$$\sum_i x_i = T \tag{7.4}$$

$$\min(\mu_{\tilde{A}_i}(x_i)) = \text{as high as possible, } \forall i \tag{7.5}$$

This not necessarily would result in a unique solution: increasing one x_i, while decreasing another $x_j, j \neq i$, might still keep the conditions satisfied and both solutions can have the same lowest membership grade among the involved fuzzy sets. To overcome the problem that the solution is not necessarily unique, an additional criterion will be introduced. The algorithm starts from a known defuzzifier; the differences between the defuzzified value of each fuzzy set and the initial value of chosen defuzzifier for this fuzzy set should be as small as possible:

$$abs(x_i - \Delta_i) = \text{as low as possible, } \forall i \tag{7.6}$$

where x_i is the final defuzzified value for fuzzy set $\tilde{A}_i$ and $Delta_i$ is the value of the initial defuzzifier for fuzzy set $\tilde{A}_i$.

In this case, we aim to keep the values as closest to the MeanOfMax as possible. The reason for this choice stems from the fact that the fuzzy sets in our problem are convex and experiments had shown that the MeanOfMax already yields quite a good approximation of the outcome.

The algorithm to achieve this starts by finding the height of each fuzzy set, and checks if the sum can be achieved using values from the alpha levels of the fuzzy sets at their heights. If not, lower alpha-levels are used to define new intervals which are tried until intervals are found from which values that sum up to the total are found.

The developed algorithm is illustrated in pseudo-code in Algorithm 4. The first 7 lines help determine the starting point: for each fuzzy set $(\tilde{A})_i$, the MeanOfMax (m_i), height (h_i) and the weak alpha cut $([c_i^l, c_i^r])$ at the height are calculated.

Algorithm 4 Simultaneous defuzzification using shifted Mean of max

---

1: **for all** i **do**
2:    $m_i \leftarrow \text{MeOM}(\tilde{A}_i)$
3:    $h_i \leftarrow \text{height}(\tilde{A}_i)$
4:   ▷ highest occurring membership grade
5:    $[c_i^l, c_i^r] \leftarrow \tilde{A}_{i_{h(A)_i}}$
6:   ▷ reverts to core for normalized $\tilde{A}_i$
7: **end for**
8: **if** $T \in \sum_i [c_i^l, c_i^r]$ **then**
9:    **for all** j **do**
10:      $x_j \leftarrow m_j + (T - \sum_i m_i) \frac{c_j^r - c_j^l}{\sum_i (c_i^r - c_i^l)}$
11:    **end for**
12: **else**
13:    $a_i^l \leftarrow c_i^l, a_i^r \leftarrow c_i^r$
14:    **if** $T < \sum_i c_i^l$ **then**
15:      **repeat**
16:        $c_i^l \leftarrow a_i^l$
17:        $\alpha \leftarrow \text{findNextAlphaLeft}(A_i)$
18:        $]a_i^l, a_i^r[= A_{i_{\widetilde{\alpha}}}$
19:      **until** $T \in \sum_i]a_i^l, c_i^l[$
20:                                   ▷ Left shifts more left / Right becomes previous left
21:      **for all** j **do**
22:        $x_j \leftarrow c_j^l - (\sum_i c_i^l - T) \frac{a_j^l - c_j^l}{\sum_i (a_i^l - c_i^l)}$
23:      **end for**
24:    **else if** $T > \sum_i cr_i$ **then**
25:      **repeat**
26:        $c_i^r \leftarrow a_i^r$
27:        $\alpha \leftarrow \text{findNextAlphaRight}(A_i)$
28:        $]a_i^l, a_i^r[= A_{i_{\widetilde{\alpha}}}$
29:      **until** $T \in \sum_i]c_i^r, a_i^r[$
30:                                   ▷ Right shifts more right / Left becomes previous right
31:      **for all** j **do**
32:        $x_j \leftarrow c_j^r + (T - \sum_i c_i^l) \frac{a_j^r - c_j^r}{\sum_i (a_i^r - c_i^r)}$
33:      **end for**
34:    **end if**
35: **end if**

If the constraining value T is an element of $\sum_i [c_i^l, c_i^r]$, the corrected defuzzification for each fuzzy set can be immediately calculated by the formula:

$$x_j \leftarrow m_j + (T - \sum_i m_i) \frac{c_j^r - c_j^l}{\sum_i (c_i^r - c_i^l)} \qquad (7.7)$$

In this case, the lowest occurring membership grade will be the lowest height.

If however the constraining value T is not an element of $\sum_i [c_i^l, c_i^r]$, it means that it is not possible to satisfy the constraint while the lowest occurring membership grade will be the lowest height. Lowering the alpha-level allows us to apply the algorithm on bigger intervals, which is what happens in the next steps. Notice that, the fuzzy sets are assumed to be convex, all α-levels are intervals.

The functions *findNextAlphaLeft* and *findNextAlphaRight* look for the next α-level. This can be a simple lowering of the α level with a fixed amount. In our implementation, where piecewise linear fuzzy sets are used, this function searches for the first breakpoint with a membership grade smaller than the current α level. The membership grade of this next breakpoint provides the next α level. The former will be used if the constraining value T is smaller than $\sum_i c_i^l$, the latter if T is greater than $\sum_i c_i^r$.

From the construction, the algorithm maximizes the lowest membership grade and yields the solution with the shortest distances to MeanOfMax. It satisfies the aforementioned criteria as follows:

1. Core selection: the algorithm will select elements from the core, if the constraint allows for this. In other words: if the given value T that constrains the sum of the defuzzified values is obtainable from values in the cores of the fuzzy set, all returned defuzzified values will belong the cores of their respective fuzzy sets.
2. Scale invariance: the algorithm exhibits scale invariance. Ratio scale is satisfied provided all fuzzy sets and the constraining value undergo the transformation. Relative scale is satisfied when the constraining value undergoes the sum of all the translations of the fuzzy sets involved. Interval scaling is satisfied under the combination of conditions for ratio and relative scale. This is explained through the fact that MeanOfMax exhibits scale invariance, and the operations in the algorithm are not scale dependent.
3. Monotony: if one fuzzy set is replaced by a fuzzy set that represents bigger values, while all other sets and the constraining value remain the same, monotony is not guaranteed. This is due to the fact that the values are adjusted in the last stage, and the amount of adjustment for the new fuzzy set can differ. Even if the constraining value also changes, it remains impossible to predict how this will affect the defuzzified value of one fuzzy set without further details on the shape of the fuzzy sets.
4. x-Translation: the algorithm satisfies x-Translation if the constraining value is translated over the sum of all the translations of the different fuzzy sets.
5. x-Scale: the algorithm satisfied x-Scale if the fuzzy sets and the constraining value are scaled with the same amount.

7.2.3 Examples

Consider the fuzzy sets A, B, C and D, with membership functions defined below and shown on Fig. 7.3.

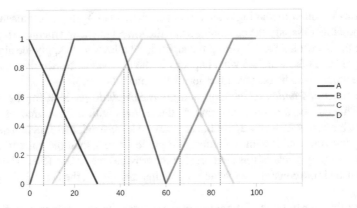

Fig. 7.3 Example fuzzy sets A, B, C and D to illustrate the constraint defuzzification

$$\mu_{\tilde{A}} : \mathbb{R} \to [0, 1]$$

$$\mu_{\tilde{A}}(x) = \begin{cases} \frac{-1}{30}x + 1 \; \forall x \in [0, 30] \\ \quad 0 \qquad \text{elsewhere} \end{cases}$$

$$\mu_{\tilde{B}} : \mathbb{R} \to [0, 1]$$

$$\mu_{\tilde{B}}(x) = \begin{cases} \frac{1}{20}x & \forall x \in [0, 20[\\ \quad 1 & \forall x \in [20, 40] \\ \frac{-1}{20}(x - 40) + 1 & \forall x \in [40, 60] \\ \quad 0 & \text{elsewhere} \end{cases}$$

$$\mu_{\tilde{C}} : \mathbb{R} \to [0, 1]$$

$$\mu_{\tilde{C}}(x) = \begin{cases} \frac{1}{40}(x - 10) & \forall x \in [10, 50[\\ \quad 1 & \forall x \in [50, 60] \\ \frac{-0.2}{20}(x - 60) + 1 & \forall x \in]60, 80] \\ \frac{-0.8}{10}(x - 80) + 0.8 & \forall x \in]80, 90] \\ \quad 0 & \text{elsewhere} \end{cases}$$

$$\mu_{\tilde{C}} : \mathbb{R} \to [0, 1]$$

$$\mu_{\tilde{C}}(x) = \begin{cases} \frac{1}{30}(x - 60) \; \forall x \in [60, 90[\\ \quad 1 \qquad \forall x \in [90, 100] \end{cases}$$

The algorithm works based on α levels at breakpoints. Due to the definition as convex fuzzy sets, all α levels of the sets in the example are intervals. The Table 7.1 shows a number of these intervals for given α values. In a constraint defuzzification, it is obvious that situations can occur where the constraints are too limiting and would prevent any solution to be found. The last column in Table 7.1 lists the supports of the involved fuzzy sets. The sum of the supports is [70, 280], implying that if

Table 7.1 Intervals obtained from the fuzzy sets A, B, C and D

	kernel	$\alpha = 0.8$	support
$\tilde{A}$	[0, 0]	[0, 6]	[0, 30]
$\tilde{B}$	[20, 40]	[16, 44]	[0, 60]
$\tilde{C}$	[50, 60]	[42, 80]	[10, 90]
$\tilde{D}$	[90, 100]	[84, 100]	[60, 100]
Sum	[160, 200]	[142, 230]	[70, 280]
Center	180	186	175

the desired sum of defuzzified values is outside of this interval, no solution will be found. Opposite, the sum of the kernels-intervals is [160, 200]. As the algorithm aims at maximizing the value of the lowest membership grade, it should result in satisfying a constraint that sums up to a value in this interval while keeping the lowest membership equal to 1.

To illustrate this, consider that the sum has to be 200. The centers of the core intervals sum up to: $0 + 30 + 55 + 95 = 180$. The difference with the target value is 20. The lengths of the intervals are 0, 20, 10, 10, with the total length of all core intervals equalling 40. Following the algorithm, the values are adjusted as follows:

$$x_A = 0 + (200 - 180)\frac{0}{40} = 0$$

$$x_B = 30 + (200 - 180)\frac{20}{40} = 30 + 10$$

$$x_C = 55 + (200 - 180)\frac{10}{40} = 55 + 5$$

$$x_D = 95 + (200 - 180)\frac{10}{40} = 95 + 5$$

The biggest intervals values are adjusted the most; the lowest occurring membership grade is 1. The calculations are similar for a constraining value of 160, and yield the smallest element of each core interval. In between, all elements values are adjusted linear in accordance with their length.

Consider now that the constrained value is decreased to 159; this constraint can no longer be solved using only values for the cores. The algorithm requires lowering of the α-level value. The function *findNextAlphaLevelLeft* scans from membership grade 1 (the starting value) down to 0, only considering the left sides of the fuzzy sets, and stops when it encounters a breakpoint, or reaches 0. In this case, 0 will be reached: the intervals in the column *support* in Table 7.1 will be used; the formula in the algorithm (line 18) yields:

$$x_A = 0 - (160 - 159)\frac{0 - 0}{70 - 160} = 0$$

$$x_B = 20 - (160 - 159)\frac{0 - 20}{70 - 160} = 19.78$$

$$x_C = 50 - (160 - 159)\frac{10 - 50}{70 - 160} = 49.56$$

$$x_D = 90 - (160 - 159)\frac{60 - 90}{70 - 160} = 89.67$$

The defuzzified values are such that the smallest membership grade of each of the values in their respective fuzzy set is maximized. The amount of adjustment is inverse proportional to the steepness of the left side of the fuzzy set: B, which has the steepest left side adjusts the least; D, which has the least steep left side adjusts the most. The values continue to decrease linearly as the constraining value T is decreased; all the way till values match the minimum possible values of the respective supports.

The algorithm behaves similarly on the right side of the fuzzy sets. To illustrate the need for finding the breakpoints, consider the example where the value $T = 229$. This situation is similar to the previous one, with the difference that *getNextAlphaLevel-Right* will return 0.8, as there is a breakpoint in C. The algorithm yields:

$$x_A = 0 + (229 - 200)\frac{6 - 0}{230 - 200} = 5.8$$

$$x_B = 40 + (229 - 200)\frac{44 - 40}{230 - 200} = 43.87$$

$$x_C = 60 + (229 - 200)\frac{80 - 60}{230 - 200} = 79.33$$

$$x_D = 100 + (229 - 200)\frac{100 - 100}{230 - 200} = 100$$

In this example, the term added by C is the largest, as it is the least steep fuzzy set. This changes when the constraining value increases further; for $T = 231$, the result becomes:

$$x_A = 6 + (231 - 230)\frac{30 - 6}{280 - 230} = 6.48$$

$$x_B = 44 + (231 - 230)\frac{60 - 44}{280 - 230} = 44.32$$

$$x_C = 80 + (231 - 230)\frac{90 - 80}{280 - 230} = 80.2$$

$$x_D = 100 + (231 - 230)\frac{100 - 100}{280 - 230} = 100$$

This time, the term added by C is the smallest, as now the right side of this fuzzy set is the steepest. The algorithm manages to balance the terms in order to maximize the lowest membership grade.

It is obvious that the algorithm is only applicable if the fuzzy sets are convex. The above calculations use the fact the fuzzy sets are continuous and piecewise linear, however the concept of the approach is more general and can be extended to suit non-linear and even non-continuous fuzzy sets.

Chapter 8
Data Comparison

Abstract In the development of the artificial intelligent system for processing spatial data, several side problems appeared. These problems are not necessarily directly related to the main problem, but also needed to be solved in order to either evaluate or implement the algorithm. The first developed aspect is an algorithm for assessing the similarity of a grid with a given reference grid, based on how well the grid resembles the unknown underlying distribution. This approach is used in the algorithm to select parameters, but is also used to evaluate the results of different algorithms. The second development concerns a way of dealing with robustness errors when processing spatial data. The limited representation of real numbers in a computer system causes rounding errors, which—if this happens for numbers that represent coordinates— may result in robustness errors concerning the geometry calculations. The proposed method aims to find errors and circumvents them by lowering the dimension of calculated geometries. This algorithm is used throughout the implementation, as all spatial operations make use of it.

8.1 Introduction

Gridded data is commonly an approximation, in the form of a discretization, of a property that has differing values over the region of interest. As such, there is an underlying distribution, but this distribution is not known and not captured in the representation of the raster. As different assumptions can be made regarding the underlying spatial distribution to perform a regridding, multiple rasters of the same resolution and grid layout can be calculated to represent the same data, but they could still differ in the values for individual cells. As such, particularly when developing new methods for regridding, it is necessary to assess which of these solutions is the best one in order to assess the quality of different methods. Such an assessment can also help in determining the ideal parameters for use in the developed algorithm. Traditionally, methods to compare datasets such as correlation are used. However, as will be shown by means of small examples, these methods tend to ignore the spatial aspects and merely compare the values. A new method is developed to compare gridded datasets, in a way that takes into account the spatial aspect in order to detect

© Springer Nature Switzerland AG 2019

J. Verstraete, *Artificial Intelligent Methods for Handling Spatial Data*, Studies in Fuzziness and Soft Computing 370, https://doi.org/10.1007/978-3-030-00238-1_8

to what extent the datasets are approximating the same underlying distribution. A reference dataset is generated, against which the different datasets are matched; this results in a value indicating how well their spatial distribution resembles that of the reference dataset.

8.2   Current Approaches

8.2.1   Case Study

To illustrate the downsides of the current ranking methods used, consider three grids: X, Y and Z as on Fig. 8.1. All three grids consist of n cells, perfectly aligned and all representing the underlying distribution of a line. Grid X is considered to be the reference grid, and we need to assess which of the grids Y and Z best resembles the grid X. Intuitively, grid Z better resembles the line than grid Y, particularly as the only knowledge available is grid X, and Z better concentrates the values around the same cells. This is an abstract example, but it serves as an illustration for the problem, and will be used as example in this section.

8.2.2   Value Comparison: Typical Approaches

Correlation

Grids can be seen as finite sets of values, and as a result Pearson correlation is often used to compare sets. As the grids are defined on the same raster, they have the same number of elements. Pearsons's product-moment correlation between a list of values A and a list of values B is defined as [2]:

Fig. 8.1 Case study: which grid resembles grid X best: grid Y or grid Z?

0	0	0	0
100	100	100	100
0	0	0	0
0	0	0	0

Grid X

0	0	0	0	20	20	20	20
50	50	50	50	60	60	60	60
50	50	50	50	20	20	20	20
0	0	0	0	0	0	0	0

Grid Y   Grid Z

$$corr(A, B) = \sum_{i=1}^{n} \frac{(a_i - E(A))(b_i - E(B))}{(n - 1)s(A)s(B)} \tag{8.1}$$

$E()$ is the expected value, approximated by the mean of the values and $s()$ is the notation for the standard deviation of the values. The values a_i and b_i are the i-th values in respectively population A and population B. Positive values indicate a proportional correlation (higher values in A relate to higher values in B); negative values imply an inverse one. In the grid context; the reference grid is one population (e.g. A) and a grid to be ranked takes the role of the second population; a higher correlation indicates a higher similarity. Using formula (8.1) on the above example with X as the reference results in $corr(X, Y) = 0.577$, while $corr(X, Z) = 0.927$; indicating that when using the Pearson correlation, Z resembles X better than Y does.

Sum of Squared Differences

The sum of squared differences (SSD) is a commonly used method to compare differences between ranges of values. As the name implies, for every cell, the difference of the values is squared, and the resulting values for all cells are added up:

$$SSD(A, B) = \sum_{i=1}^{n} (a_i - b_i)^2 \tag{8.2}$$

As before, A and B represent two lists of values, and a_i respectively b_i the i-th value in A and B. The smaller the sum of squared differences, the more both populations are similar; as before the cell values of the reference grid form one population, those of a grid to be ranked the other. Using the above example, the SSD behaves similarly as correlation and considers that Z is better than Y: $SSD(X, Y) = 20000$ is greater than $SSD(X, Z) = 9600$.

Sum of Weighted Differences

A modification to the method of least squares is obtained by the introduction of weights: every term i of the sum is multiplied with a weight w_i. This weight can be introduced to increase the impact of this particular term. One example in the context of grids would be to increase the weights for grid cells with higher values, to indicate that any difference in those is counted stronger than for other cells.

$$S(A, B) = \sum_{i=1}^{n} w_i (a_i - b_i)^2 \tag{8.3}$$

The notations for A, B, a_i and b_i are the same as before. There is however no justification for giving a higher weight to specific values; in general the behaviour is similar as for the sum of squared differences.

Fig. 8.2 Example with 3 cells

A	B	C

Table 8.1 Table comparing the performance of correlation, sum of squared differences and sum of weighted squared differences

	Ideal	Examples							
Cell	Value	1	2	3	4	5	6	7	8
A	100	90	80	70	60	60	50	50	34
B	0	10	10	20	30	20	40	30	33
C	0	0	10	10	10	20	10	20	33
Correlation		0.995	1	0.987	0.917	1	0.693	0.945	1
SSD		200	600	1400	2600	2400	4200	3800	6534
SWSD		300	1000	2300	4200	4000	6700	6300	10890
SWSD1		1300	3700	8800	16500	14800	26800	23700	40293
SWSD2		1800	5000	12000	22600	20000	36800	32200	54450

8.2.2.1 Discussion

To better show the shortcomings of the above methods, a different example is used. Consider a grid of just three cells, as in Fig. 8.2; the cells are labeled A, B and C. Table 8.1, lists ideal solution, which is considered the reference, as well as eight possible other grids that need to be ranked. The correlation, sum of squared differences and sum of weighted squared differences are listed as well. For the latter, two different definitions for the weights were considered (SWSD1 and SWSD2), using the distance between cells as an attempt to include the spatial aspect. In SWSD1, the weight for a cell itself is 3, the weight for a neighbouring cell 2, and the weight for a cell further away 1. For A, this means that the weight for A is 3, for B 2 and for C 1; for B it means that B has a weight of 3 and both A and C have a weight of 2. The case for C is similar to A. In SWSD2, the weights decrease with quadratic distance; the weight for the cell itself is 4, for the neighbouring cells 3 and for the furthest cells 1.

The Pearson Correlation ranks options 2, 5 and 8 as equal and as the best, as they exhibit the same pattern of values: one higher value for A and two equal lower values for B and C. The spatial aspect should favour the patterns where, if values for B are lower than those of A they should still be higher than those of C. In the sum of squared differences (SSD), lower values imply a better ranking and it ranks for example 5 above as 4, as biggest difference is smaller regardless of the cells where this deviation occurs. Adding weights (SSWD1 and SSWD2) can change the situation somewhat but there is no real justification for those weights. Still, option 4 is ranked below 5 and option 6 below 7, contrary to intuition. In a more realistic example, there will be multiple cells at the same distance, and in this case the change of one value can be compensated by the opposite change of a value of a cell at the same distance, which would not be detected using SSD.

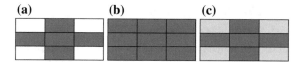

Fig. 8.3 Three examples of possible masks, with the considered cells shaded in grey: direct neighbours (**a**), all neighbours (**b**) or a weighted combination where the shade of grey is indicative for the weight of the cell (**c**)

Intuitively, it is clear that option 1 is a very good approximation. Ranking the options 5 and 6 is more difficult: option 5 has a higher value for A than option 6, which would make it better, but at the same time it has a more spread out distribution over the cells B and C, which is should rank it lower. The option 8 clearly is the worst of the considered options, as it spreads out the data almost evenly over all three cells and does not resemble the real data.

8.3 Developed Fuzzy Set Approach

8.3.1 Preprocessing of the Reference

The developed approach starts from preprocessing the reference grid: for each cell, a fuzzy set will be determined. This fuzzy set holds possible values for the cell and has the reference value with membership 1 while the shape of the set is determined using neighbouring cells. As such, this fuzzy set will incorporate the spatial aspect. How good a grid matches will be determined by comparing its values with these fuzzy sets.

To determine the fuzzy sets, the values of the neighbouring cells are needed, but there are different definitions possible on what constitutes the neighbouring cells of a given cell. In a general case of irregular grids, one could consider all cells within a certain distance of the given cell. For regular grids, a mask can be applied. A mask is a form of sliding window that defines which cells are considered; on Fig. 8.3a a mask that defines the direct neighbours is shown, but it is also possible to define all eight neighbours (Fig. 8.3b) or even consider a mask that involves weights, as illustrated by lighter shade on Fig. 8.3c. The choice of mask can be used to give preference to specific underlying patterns, offering additional possibilities of fine-tuning the ranking.

8.3.2 Defining the Ranking Value

With each cell of the reference grid, a fuzzy set will be associated that represents the
values that are deemed suitable for this location. This set will have the ideal value
at the core. In [43], the details on the construction of the fuzzy sets are provided.
First, it is assumed that the possible values of a cell are considered to be between the
minimum of the values covered by the mask L_i, whereas the maximum is considered
to be the total of all cells covered by the mask U_i, yielding the interval $[L_i, U_i]$
for cell i. Using this interval as the core, a triangular fuzzy set can be constructed.
However, there is a possibility that this fuzzy set is skewed, in which case a small
deviation from the ideal value would not have a symmetrical effect, which may
incorrectly favour either higher or lower values. To compensate, the support is defined
using $\max(f(a_i) - L_i, U_i - f(a_i))$, where $f(a_i)$ is the ideal value of cell i in the
reference grid A, resulting in the interval $[x_1^i, x_2^i]$ with x_j^i defined as

$$x_1^i = f(a_i) - \max(f(a_i) - L_i, U_i - f(a_i))$$
$$x_2^i = f(a_i) + \max(f(a_i) - L_i, U_i - f(a_i)) \tag{8.4}$$

These limits are used to define the functions f_1^i and f_2^i that make up respectively the
left and right side of the triangular fuzzy set for cell a_i:

$$f_1^i(x) = \frac{1}{f(c_i) - x_1^i}(x - f(c_i)) + 1$$
$$f_2^i(x) = \frac{1}{f(c_i) - x_2^i}(x - f(c_i)) + 1 \tag{8.5}$$

As negative values are not allowed (in this case), the values of the grid should not
be negative, so the support is restricted to positive values. For every cell a_i of the
reference grid A, the fuzzy set is defined by the membership function

$$\mu_{\tilde{A}}^F : \mathbb{R} \to [0, 1]$$
$$x \mapsto \mu_{\tilde{A}}(x)$$

where

$$\mu_{\tilde{A}}^F(x) = \begin{cases} f_1^i(x) & \text{if } x \in [\max(0, x_1^i), f(c_i)] \\ f_2^i(x) & \text{if } x \in [f(c_i), x_2^i] \\ 0 & \text{elsewhere} \end{cases} \tag{8.6}$$

These fuzzy sets allow every cell of a candidate grid to be matched with the
relating cell in the reference grid. This yields a value in the range [0, 1], indicating
how well the cell matches with the reference cell.

The values for the different cells need to be aggregated in order to yield a single
value that reflects how well the grid matches. For this aggregation, we propose to
use the average, as this allows for worse matching cells to be compensated by better
matching cells. As such, the value assigned to a given grid B is calculated by:

$$m_A(B) = \sum_{i=1}^{n} \frac{1}{n} \mu_{\tilde{C}_i^A}^A (b(c_i)) \tag{8.7}$$

where c_i, $i = 1..n$, are the cells of the raster, $b(c_i)$ are the cell values of grid B, $\tilde{C}_i^A$ is the fuzzy set derived for cell c_i that represents which values are good approximations for the cell in grid A and $\mu_{\tilde{C}_i^A}^A$ is its membership function.

The function is non-commutative, which can be shown by a simple example; hence the choice for the notation $m_A(B)$ over $m(F, G)$. In addition, if B_1 and B_2 are two different grids, the knowledge that $m_A(B_1) < m_A(B_2)$ does not make it possible to make statements on how $m_{B_1}(B_2)$ relates to $m_{B_2}(B_1)$.

As the membership function is symmetrical, the impact of a cell c_i whose value is $b(c_i) = a(c_i) - \epsilon$ is the same as $b(c_i) = a(c_i) + \epsilon$, for all possible epsilons greater than $a(c_i)$. The function is also increasing: if the values of a grid B are closer to those of the reference grid than the values of a grid C, than it will be ranked better.

In [43], the developed ranking function is further illustrated by means of examples on artificial grids. These examples show that the function ranks based on spatial distribution rather than blind comparison of the occurring values.

Chapter 9
Algorithm

Abstract This chapter contains the description of the developed algorithm, applying the different aspects considered in the earlier chapters.

9.1 Data

The developed algorithm will first be illustrated using the spatial disaggregation problem. The reason is that the problem of regridding can be considered a special case of a spatial disaggregation problem when introducing the segment grid as introduced in Sect. 3.1.2. The problem of data fusion is slightly different, and will only partly be considered.

The algorithm for spatial disaggregation starts with identifying suitable parameters. This is done by exhaustively calculating the value and an associated range for each of the output cells (Sects. 5.1 and 5.2). The range is used to rescale the values, which are assigned to a new grid that has the same layout as the output grid. The similarity of this grid to the output grid can be used to assess how good the parameter is: a good parameter will result in a grid that resembles the output grid. This is only the case for parameters with a positive correlation, however the rescaling operation can be done in such a way that the evaluation will hold also for parameters with a negative correlation. This selection procedure is described in Sect. 5.3. For this similarity, the method developed in Chap. 8. This process is also performed to determine which output parameter is the most suitable: this parameter and the most suitable range for it need to be calculated when the rulebase will be applied.

The best parameter/range combinations for input values will be used in the rulebase construction. The number of parameters is chosen beforehand; in addition a threshold can be imposed on the match value of the parameter to further limit the selection of less suitable parameters. For solving a regridding problem, the grid OUTPUT is replaced by the segment grid as defined in Sect. 3.1.2. For a data-fusion problem, there is no input grid, so only parameter and range definitions that make no use of an input grid can be considered.

© Springer Nature Switzerland AG 2019
J. Verstraete, *Artificial Intelligent Methods for Handling Spatial Data*, Studies in Fuzziness and Soft Computing 370, https://doi.org/10.1007/978-3-030-00238-1_9

9.2   Constructing the Rulebase

To construct the rulebase, a training set is needed. In the case of a disaggregation or a data fusion problem, this can be available and used. It is however unlikely that there is a training set for regridding, as the segment grid is very specific to the combination of input and output grid. From experiments it was observed that the relative layouts of the grids are enough to find suitable parameters; as such is it possible based on the input, output and proxy data to generate a training set. In Chap. 10, the method for generating trainingsets in the experiments is presented. A trainingset can also be generated for spatial disaggregation or datafusion. The rulebase construction makes use of the developed algorithm for constructing rulebases with variable spaces, as developed in Sect. 6.4. For the algorithm, it is assumed that we have a list of suitable parameters: this is a combination of a definition for a value and an associated range.

With a list of value-definitions, ranges and a training set, the algorithm to construct rules using variable spaces described in Sect. 6.4 can be applied.

Algorithm 5 Construction of the rulebase

1: Determine list of parameters
2: Determine linguistic term distributions for parameters
3: **for all** each output grid cell **do**
4:     **for all** parameter **do**
5:         calculate value
6:         calculate minimum possible
7:         calculate maximum possible
8:         map linguistic term distribution on min-max range
9:         find best matching linguistic term with given value
10:     **end for**
11:     make rule using parameters/linguistic terms
12:     assign rule lowest occurring membership grade as weight
13:     **if** no rule with parameter/linguistic term exists **then**
14:         replace rule with new rule
15:     **end if**
16:     **if** weight of existing rule < weight of new rule **then**
17:         replace rule with new rule
18:     **end if**
19: **end for**

9.3   Applying the Rulebase and Defuzzifying the Results

The application of the rulebase has to take into account the fact that the rulebase uses variable spaces. The modification to the standard rulebase application (Chap. 6) needs to calculate the range when it is needed, which is at the same time as when the value is calculated. This modification is listed in the pseudocode below.

Algorithm 6 Application of the rulebase

1: **for all** each output cell **do**
2:    **for all** each parameter **do**
3:       calculate parameter value
4:       calculate parameter ranges
5:       map linguistic term distribution on range
6:    **end for**
7:    use parameter values in the rulebase
8:    output cell ← evaluate(rulebase)
9: **end for**

The above pseudocode results in a fuzzy set for every output cell, for the defuzzification of the fuzzy sets the methodology developed in Chap. 7 will be applied:

Algorithm 7 Defuzzification of the result

1: **for all** each input cell **do**
2:    Find all target cells that overlap with input cell
3:    value of input cell provides the shared constraint
4:    use constraint defuzzification on the fuzzysets of these target cells
5: **end for**

The *constraint of the input cell* depends on the problem considered: for a property that is an absolute value, this constraint is that the sum of the defuzzified values has to add up to the value of the input cell. However, for other properties, this constraint can be different (average, weighted average, etc.). The defuzzification results in the final values. When the rulebase approach is used to solve a general regridding problem, this end result is the segment grid. To present it as the original output, the segments have to be combined to form the cells of the original output grid.

9.4 Complexity

The complexity of the algorithm mainly depends on the operators used to determine the parameters. The construction of the rulebase requires calculating all the parameters for each of the output cells. This is linear in the number of output cells, if the calculation of a parameter (the value or the most possible range) is not dependent on the number of output cells. The only situation in which this occurs is if the most possible range is a global range, but in this case the complexity can be controlled by performing this calculation only once and using this cached value whenever it is needed. As such, the complexity of the construction of the rulebase is linear.

The rulebase is evaluated once for every cell in the output grid. The size of the rulebase is not dependent on the output grid. Consequently, the complexity of its application also depends on the calculation of the parameters. The same remark as for the construction of the rulebase holds, making the evaluation of the rulebase of linear complexity in terms of the number of output cells.

As a last step, the defuzzification is necessary for all output cells, but the complexity of the defuzzification a single cell is not dependent on the size of the output grid. As such, the final defuzzification step is also of linear complexity.

Part III
Algorithm and Experiments

The last part details on aspects of the implementation and experiments. This includes issues encountered and programmatic choices made in order to implement the algorithm. It also details the environment in which experiments were performed; this includes a presentation of the test data, an overview of the test cases and their purpose and the results of the experiments. The part ends with a conclusion, in which the novel developments needed to create the methodology are summarized and an overview for future research and other possible applications are presented.

Part III
Algorithm and Proof in Ethics

Chapter 10
Implementation Aspects

Abstract In the development of the artificial intelligent system for processing spatial data, several side problems needed to be solved. These problems are not necessarily directly related to the main problem, but also needed to be solved in order to either evaluate or implement the algorithm. Among other things, this includes the way of dealing with robustness errors when processing spatial data. The limited representation of real numbers in a computer system causes rounding errors, which—if this happens for numbers that represent coordinates—may result in robustness errors concerning the geometry calculations. The proposed method aims to find errors and circumvents them by lowering the dimension of calculated geometries. This algorithm is used throughout the implementation, as all spatial operations make use of it.

10.1 Programatory Aspects

In order to verify the developed algorithm, an implementation was required. This implementation was developed in Java 8 [1] and runs both on Windows as well as Linux. For specific spatial functionality, the JTS Spatial Library [21] was used. In addition to the JTS Library, several classes from the JUMP Framework [22] were also used as building blocks for geographic operations. The Apache Commons Math library [3] was used for basic mathematical functions such as the Pearson correlation. For the operations involving fuzzy sets, initially jFuzzyLogic [19] provided the internal representation but recently this was changed to library FuzzyJ [30]. The experiments performed here used FuzzyJ; the change was made to be able to use additional functionality that is available in FuzzyJ.

The current implementation is mainly intended as a proof-of-concept and prototype; the raster data is represented as a set of features, where each feature represents a pixel of a gridded dataset (this model is described in Sect. 1.2.3). While many parts of the algorithm can be performed in parallel, the implementation is not multi-threaded. The main reason for this was to keep the complexity of the implementation down. Priority was given to a modular approach where many different aspects of the

© Springer Nature Switzerland AG 2019
J. Verstraete, *Artificial Intelligent Methods for Handling Spatial Data*, Studies in Fuzziness and Soft Computing 370, https://doi.org/10.1007/978-3-030-00238-1_10

algorithm (rulebase construction approaches, different parameters, various defuzzi-fication methods) could be easily swapped out or reconfigured.

An implementation has to cope with limitations of the platform; when processing spatial data, and then in particular geometric aspects of the spatial data, the internal computer representation of the data poses specific challenges due to the limited accuracy of the representation of real numbers in a computer system. These challenges, and the way they were handled are described in Sect. 10.2. The specific approach to implementing the different parameters in a modular way is described in Sect. 10.3. The algorithm needed to be verified for correctness as well as accuracy. To achieve this, specific datasets were generated as explained in Sect. 11.1. Lastly, the algorithm uses training data in order to generate the rulebase system. Due to the specific nature of the problem, it is possible to generate a training set for a given case, which increases the applicability; the methodology to do this is explained in Sect. 10.4.

10.2 Geometry Calculations

10.2.1 Introduction

Processing spatial data requires a number of geometric computations; in the considered problem of overlaying data, there is the determination of overlap, intersection and distance, all of which are important to identify which cells relate to a given cell. Geo-referenced data adds numerous computations that relate to transforming data a given coordinate system and projection system. The limited representation of floating point numbers in a computer system can cause the results of operations not to be representable in the computer system. For a detailed overview of a classification of these problems and causes, we refer to [15], Chap. 4. The problem of robustness often calls for solutions tailored to the application. A custom solution to deal with it in the context of regridding and spatial disaggregation was developed as the functionality present in the JTS framework proved insufficient. This developed solution is presented in this section.

First, it is necessary to consider what robustness problems are and what causes them. To illustrate, consider the example in Fig. 10.1. On the Figure, two lines are shown: line A connects $(1, 4)$ with $(4, 0)$; line B connects the points $(2, 1)$ and $(4, 4)$. If we assume that it is only possible to use integer values as coordinates then the intersection point of the lines, indicated by the circle, cannot be represented. The closest point that can be represented is the point $(3, 2)$, indicated by the dot. The same happens on a computer system using the floating point representation, just at a higher level of accuracy. The outcome is that the mathematical result of a calculation cannot always be represented in the system, in turn resulting in the possibility that the intersection point of two lines does not belong to one or even either of the lines.

This rounding of coordinates also affects other geometries, and as such impacts e.g. intersection calculations. This is illustrated on Fig. 10.2. Figure 10.2a shows two

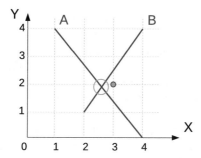

Fig. 10.1 Example of the problem of limited representation. The coordinates of the intersection point would get rounded, resulting in a point that no longer coincides with either line

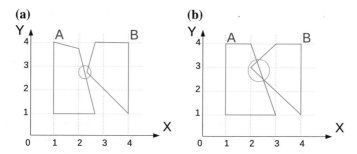

Fig. 10.2 Example of false positive intersections due to rounding of coordinates. The polygons A and B in (**a**) do not intersect, but when their coordinates are rounded (**b**), they have an intersection

non-intersection polygons *A* and *B*, but due to the rounding of their coordinates to the nearest integer, the polygons will be considered to intersect, as on Fig. 10.2b. In reality, the coordinates will not be rounded to the nearest integer, but will be limited by the internal number representation. As the coordinates most likely are the result of a coordinate transformation to perform the georefencing, this internal representation can be a limiting factor.

Furthermore, the use of different algorithms can yield contradictory outcomes: determining if there is an intersection does not necessarily requires calculating it. When such errors happen in general, geometric calculations may yield the wrong result: a cell of a grid may be considered a neighbour when it is not, there would be an intersection when there should not be, or it is even possible that one algorithm says there is an intersection while its calculation does not exist.

In the JTS Spatial library, [21], the standard mechanism to deal with robustness is to lower the precision. This makes the problem more detectable, which in turn makes it controllable. The approach effectively discretizes the continuous two dimensional space in a finite set of possible values at a lower accuracy than the internal number representation allows. All points represented in the system are then mapped to the nearest coordinate in this discrete space. While this does not solve the issue, the fact that this action is known, can make it controllable in subsequent actions. Despite this, the operations that deal with determining topological relations between cells and calculating the amount of overlap of cells still were prone to the rounding issues. For the desired application, the standard workaround proved insufficient as it is necessary to correctly asses the intersection of cells and the size of the intersection.

The first goal of our approach was to make the determination of the topological relation consistent with its calculation; the second goal was to avoid false positives in these tests. In the algorithms, a false positive is worse than a false negative. To illustrate, consider the test for an overlap: if the test reveals a false positive, it means it is wrongly identified as overlapping, which may increase the impact the geometry has. A false negative on the other hand decreases the influence of a geometry that already should be very small and most likely could be ignored. As such, preventing false positives is more important for our application than avoiding false negatives. The algorithm therefor starts from intersections, and analyses if there should be an intersection or not.

10.2.2   Algorithm

The developed algorithm was first presented in [41], and constitues a method which lowers the dimension of calculated intersections, based on an automatically determined epsilon. The algorithm is a form of post-processing of the calculated intersection: if the area of a two dimensional intersection is smaller than an epsilon, the geometry is replaced by its boundary (which is a line). If the length of a one-dimensional geometry (which may be the boundary of a former two-dimensional geometry) is smaller than an epsilon, the geometry is replaced by its centroid. This action was called *collapsing* the geometry.

An important aspect in this problem is the magnitude of the values of the coordinates and how close they come to the representation limit of the system. However, it would be interesting if the system behaves the same independently of the magnitude of the coordinates. To achieve this, the epsilon value used to evaluate an intersection is calculated based on the relative size of the geometries that created it. By making the epsilon dependent on the relative size of the geometries, it helps to determine when an area (or line) is sufficiently small, relative to the areas (or lines) under consideration, to be collapsed. This increases consistency of the method, independent of units or scale used in the internal representation.

In order to achieve consistency between the detection of an overlap and its calculating, the overlap will be detected by calculating it. In order to generalize on, it was opted to change the calculation of the 9-intersection matrix for, which calculates the intersections of the interior, boundary and exterior of geometries involved. While this is less efficient, the gained consistency is necessary for the algorithm to operate properly. In the next sections, the details of applying the method for processing the 9-intersection matrix will be elaborated. This makes the results of the tests and the calculations consistent and suitable for our application.

10.2.3 Determining the 9-Intersection Matrix

Topological relations between geometries are commonly determined using the 9-intersection matrix [9]. The 9-intersection matrix contains information on the intersection of interior, boundary and exterior of two geometries. For a more formal definition of these concepts, we refer to [41]. Informally, the interior $A°$ of a geometry A is the set of all the points that belong to the region but do not belong to the boundary. The boundary ∂A consists of all the points that are neighbours of points that do not belong to the geometry. The exterior A^- contains all the points that do not belong to the geometry.

The standard 9-intersection matrix has boolean values: *true* if the intersection is not empty, *false* if it is empty. The dimensionally extended matrix holds the dimension of the intersections (false for no intersection, 0 for points or multi-points, 1 for lines or multi-lines and 2 for areas):

$$\begin{pmatrix} dim(A° \cap B°) & dim(A° \cap \partial B) & dim(A° \cap B^-) \\ dim(\partial A \cap B°) & dim(\partial A \cap \partial B) & dim(\partial A \cap B^-) \\ dim(A^- \cap B°) & dim(A^- \cap \partial B) & dim(A^- \cap B^-) \end{pmatrix} \qquad (10.1)$$

In our implementation, the intersection matrix is used with a traditional interpretation to determine the topology of two geometries. The benefit of redefining solving the robustness issue through the intersection matrix, is that all topological relations are covered and made consistent with one another.

Every possible topological relation between two geometries matches with one intersection matrix. The calculation of the matrix elements of the 9-intersection matrix will be made more robust; their meaning and interpretation stays the same, but the determination of the matrix elements accounts for the issues with rounding. For this, it had to be taken into account that in the JTS library, it is not possible to represent an interior or an exterior: the interior of a region has the same representation as the region but carries a different interpretation. The modified calculation of each element of the topology matrix is detailed in the next paragraphs.

10.2.3.1 Intersection of Interiors

The first matrix element, the intersection of the interiors of two geometries g_1 and g_2, is defined by the dimension of the collapsed intersection of the geometries if the following conditions hold:

- the collapsed intersection is not empty
- the intersection does not equal the collapsed geometry of the intersection between the boundary of g_1 and g_2
- the intersection does not equal the collapsed geometry of the intersection between g_1 and the boundary of g_2

In all other cases, the matrix element is considered false.

If the area of the intersection is small enough, it will be collapsed to a line and possibly to a point. If there was no intersection, this will remain. If there was a small intersection that got lost in rounding, it will not be found. One possible remaining problem is when the intersection between g_1 and g_2 is a line that will be collapsed to its centroid and this centroid cannot be represented. In this already unlikely scenario, it is very probable that the centroid will result into one of the defining points. However, if it would not, this causes an intersection between interiors as the interior of a point is the point. For our application, this will not raise a problem, as the intersection has no surface area. When used between segments and other data, this intersection matrix will also not cause issues in our application as the intersection between the other part of the cell and the geometry will carry the full weight.

10.2.3.2   Intersection of Interior with Boundary

The second matrix element, the intersection of the interior of g_1 with the boundary of g_2 is the dimension of the collapsed intersection between g_1 and the collapsed boundary of g_2 under the following conditions

- the collapsed intersection is not empty
- the collapsed intersection does not equal the collapsed intersection between the collapsed boundaries of both g_1 and g_2

The second condition effectively checks if the regions are not equal, as in this case there is no intersection between the interior of g_1 and the boundary of g_2. The calculation of the fourth matrix element, the intersection of the boundary of g_1 with the interior of g_2 is similar.

As the boundary cannot be an area, this matrix element is mainly important when it concerns relating the cell to other non-gridded data. Collapsing the intersection will result in the value being false if the length of the intersection is small enough. This makes an already small intersection even smaller (which leaves the rest of the geometry with the biggest weight), but also minimizes the effect of the intersection if the intersection was introduced due to the rounding.

10.2.3.3   Intersection of Interior with Exterior

The third matrix element, the intersection of the interior of g_1 with the exterior of g_2, is calculated as the dimension of the difference of g_1 and g_2 if

- the collapsed intersection does not equal the collapsed geometry of g_1

Its symmetrical counterpart, matrix element 7, is calculated similarly.

The condition verifies that both geometries are not equal, in which case the interior of one does not intersect with the exterior of the other. By collapsing the intersection, the dimension of the intersection will be decreased if the intersection is small enough. This is adequate for our application for all of the rounding cases.

10.2.3.4 Intersection of Both Boundaries

The fifth matrix element, the intersection of the boundaries of both regions, is defined as the dimension of the collapsed intersection of both boundaries. The collapsing of the intersection most likely has little effect on the end result. This matrix element is mainly of importance when it concerns the relation between cells and additional non-gridded data. Similar to the second matrix element, the behaviour of minimizing the dimension yields the desired result.

10.2.3.5 Intersection of Boundary with Exterior

The sixth matrix element considers the intersection of the boundary of g_1 with the exterior of g_2. It is defined as the dimension of the difference of the collapsed boundary of g_1 with g_2 if

- the collapsed intersection of the collapsed boundary of g_1 with g_2 does not equal the collapsed boundary of g_1

As this intersection will never be of dimension 2, it is mainly of importance when cells need to be related to additional non-gridded data. The conclusion is the same as for the above matrix element.

The eight matrix element reverses the roles of the geometries, and is calculated similarly.

Intersection of the exteriors As in the library it is not possible to define a geometry that is the size of the universe, the exteriors will always intersect with the dimension 2. This is returned as value.

10.3 Parameter Implementation

10.3.1 Values and Possible Range

The variables in the rulebase and the ranges are calculated using properties of the data and geometric properties. In the implementation, these parameters in the antecedents of the rulebase are defined using two major components: a *relates*-operator that works on a single proxy layer and a *calculator*. The *relates*-operator determines which cells of the proxy layer will be used for the calculation whereas the *calculator* performs the calculation using those cells. The relation is always considered from the point of view of an output cell and selects the cells that meet the relation specified by the *relates*-operation; the topological relations implemented are:

- cells that overlap the output cell
- cells that overlap a buffer (selectable size) around the output cell
- cells that are contained by the output cell

- cells that at a given distance (selectable) around the output cell
- cell that contains the output cell
- cells that overlap with the input cell that contains the output cell

The calculators, the component that performs an operation on the associated data of the relating cells, in this implementation are:

- sum of the values
- minimum of the values
- maximum of the values
- average of the values
- overlap-weighted sum of the values
- distance-weighted-sum of the values

By combining a *relates*-operator with a *calculator*, it is possible to specify for example that a value is *sum of all the values of cells that overlap with the output cell*. Thanks to the modular approach, the implementation can yield a large number of parameters.

For the most possible ranges, all three types considered in Sect. 5.2 were implemented using the relates operators and calculators. The distinction between all three types was maintained for performance reasons.

- global range
- zonal range
- estimated range

The global and zonal ranges both use the same calculator and the same relates operator to calculate the minimum and maximum possible value. For a global range, the values of all cells of the output are calculated and the minimum and maximum are determined and cached for performance reasons: these values are the same for all cells of the output grid and it is not necessary to recalculate them. A zonal parameter works as a global parameter but uses a limited number of cells around the considered cell rather than all cells. To achieve this, a second relates-operator is supplied: it is used to define the zone from which the output cells' values are used to determine the minimum and maximum. This area is called the zone, hence the name zonal-parameter (a global parameter is equivalent to a zonal parameter that uses the entire layer as its zone). These values are also cached. In the implementation, several definitions for this zone are considered, further increasing the number of parameters that will be considered by the system.

To define an estimated range, both the minimum and the maximum are calculated using different relates operators and value calculators. Only those combinations where the calculated minimum and the calculated maximum are suitable ranges for most of the output cells are used. The example given in Sect. 5.2.3 can by defined in terms of relates operators and calculators: for the minimum possible value, the relates operator that returns contained cells can be used in combination with a calculator that sums the values of those cells; for the maximum possible value the relates operator that returns all overlapping cells, combined with the sum of those cells can be used. In this example, the actual value of the parameter is then obtained with the same relates operator, but now computed using the area-weighted sum.

Both the input parameters and the output parameter are then defined as either a global, zonal or estimated parameter, with predetermined relates-operator(s), calculator(s) and other necessary data. In the implementation, all theoretically possible combinations are combined and tried with every proxy data. First, the most suitable output parameter is selected as explained in Part II, Sect. 5.3 using the ranking method presented in Part II, Chap. 8 and the output data. This phase requires a training set, but such a set can be generated as is shown in Sect. 10.4. For the input parameters, there is a limit on both the number of parameters as well as on the lowest ranking value that is still considered. By limiting the number of parameters, we limit the size of the rulebase.

In the current implementation, a parameter is computed using a single proxy dataset, the data of multiple proxy datasets is not combined. Ongoing research aims at considering combinations of proxy data, and at considering quantification, to allow predicates in the antecedent *if most values are high then*. Such predicates may be particularly important for data fusion.

10.3.2 Linguistic Terms

The minimum and maximum possible values define a range, which, depending on the type of parameter, may be calculated differently for each output cell. It is now still necessary to define a partitioning in linguistic terms for each parameter. In the implementation, a natural partitioning using triangular fuzzy sets is used.

For each parameter, a predefined number of linguistic terms—and thus fuzzy sets—will be used. The partitioning will be similarly defined over the different possible ranges. Evaluating a parameter therefore means first determining its value, next calculating its most possible range and finally defining the linguistic terms. In the training phase, after these three steps, the best matching linguistic term for the linguistic term can be determined. In the application phase of the rulebase, it is possible to evaluate the value against each of the linguistic terms as dictated by the rules in the rulebase system.

In the implementation, the same number of linguistic terms and a similar distribution over the most possible range was used for all parameters.

10.4 Training Data

The algorithm requires a training stage with training data: data for which an ideal solution is known. The training data should have the same grid layouts as the data that needs to be disaggregated or regridded, which may limit its existence: research that connects the proxy data to the input data may not necessarily be based on datasets with the same grid layouts as in the presented problem. This is not a problem to test the methodology, as we can generate the ideal solution in the same way as the datasets are generated, using the same underlying patterns. However, this would limit the applicability of the method. Experiments have shown that the most important factors in determining the parameters and the rulebase for a given problem are the relative grid positions. Different underlying distributions approximated with the same grid yield a nearly identical choice of parameters and constructed rulebase. This property allows us to generate a training set for any given problem: a given problem supplies the grid layouts, and these can be used to approximate synthetic datasets; this yields datasets for which the ideal output is also known. These synthetic datasets should work independent of size or resolution of the datasets and as such the way of constructing them was designed with that in mind.

To generate an underlying distribution suitable to generate a training set, the target grid is considered. This is guaranteed to be a gridded dataset, and the geometry of all cells is known. In every cell, a random number of points (0, 1, 4, 5 or 9) with random associated values are positioned. The reason to select a random number of points is to avoid that a generated dataset would have all its points on the edges of one of the datasets; the reason not to put the points totally random is to avoid extreme spatial distributions that might not be suitable, even if their likelihood of occurrence is small. By keeping the spatial distribution somewhat regulated, the system is much more likely to have the same outcomes when processing the same data. The associated values are such that the values of all points sum up to the total value of the original output grid; this guarantees that the magnitude of the values is similar. The set of all these points is an underlying distribution based on which a training set is created by sampling the points using the grid layouts from the initial problem. This process can be repeated with differently generated sets of points, consequently there is never too little training data for any given problem. An example of a generated training dataset will be shown in Sect. 11.2.

Generating a training set this way does have its limitations: the connection between the proxy data and the input data is assumed to be proportional (higher values match higher values). In general, proxy data can have many more types of connections, which will not be simulated by a training set generated using the above algorithm. This could be partly overcome by including meta-data on how given datasets relate to the input data set (proportional, inverse proportional, etc.), and generate the data accordingly.

Chapter 11
Experiments

Abstract This chapter presents experiments on various datasets using the developed method. It first introduces the origin of the test data and settings of the algorithm used in the experiments. Different experiments were performed to investigate the behaviour of the developed method with data of varying quality and geometry.

11.1 Warsaw Test Data

11.1.1 Source Data

A realistic dataset was generated in order to test the algorithm. The benefit of a generated dataset is two-fold: first, there is much more control over the data, which allows for various aspects to be tested, and second, it provides for a known ideal reference solution, which allows to judge the performance. To make the dataset realistic, it is generated using available data on Warsaw. The data at our disposal are listed in Table 11.1.

All the source data are feature-based datasets (Sect. 1.2.2). The features are thus represented by geometries, for which properties such as length (for lines) or area (for polygons) can be calculated. The geometries can be rendered using different styles; on Fig. 11.1, the dataset is shown. The road category is represented using two parallel lines (the single line that internally represents a road is drawn as two parallel lines); the different segments that make up a road are also visible. Conditional formatting allows for the different road-categories to be represented using different widths based on the associated attribute *category*; category 1 is the widest, e.g. Wybrzeże Kociuszkowskie, category 2 is medium, e.g. Aleje Jerozolimskie, whereas category 3 is the most narrow, e.g. Nowy Świat). Streets for which the attribute *restriction* indicates forbidden are shown in red (e.g. in Stare Miasto), streets which have this attribute set to restricted are shown in orange, e.g. Krakowskie Przedmieście. We noticed that some streets are wrongly labeled (e.g. some sections of Krakowskie Przedmieście carry the label *forbidden*, where it should be *restricted*), but this is not an issue when using the information to generate the test data. The figure also overlays the datasets Parks and Water; both of these sets hold polygons. The areas

© Springer Nature Switzerland AG 2019
J. Verstraete, *Artificial Intelligent Methods for Handling Spatial Data*, Studies in Fuzziness and Soft Computing 370, https://doi.org/10.1007/978-3-030-00238-1_11

Table 11.1 Datasets and their relevant attributes used for generating the Warsaw test datasets

Data	Type	Attribute	Attribute type	Content
Roads	Features	Geometry	Lines	Roads, single line representation
		Street	String	Name of the street
		Class	Integer	1: big road
				2: medium road
				3: small road
		Restriction	String	Forbidden for traffic
				Restricted traffic
Water	Features	Geometry	Polygon	Larger bodies of water e.g. Wisła
Parks	Features	Geometry	Polygon	Larger traffic-free areas e.g. Ogród Saski

that represent parks are shown in green, areas that constitute large bodies of water in blue. Both these datasets are similar in that they represent large regions where no traffic is possible; consequently they will be used as additional data to help determine the underlying distribution.

On Fig. 11.1, two sections are enlarged. The top section just misses the Old Town (it touches top right corner), and stretches to beyond the train station Warszawa Zachodnia (bottom left corner). This area was chosen as it contains a low traffic area, parks and some major roads. The bottom section highlights the area between Puławska and Jana Rodowicza Anoda. In this region, there are two lines on which traffic is concentrated (one vertical, one diagonal); this is very visible on the generated grids and this should be very visible on the outcomes of the algorithm. The representation with the two enlarged regions as shown on Fig. 11.1 will be used for all experiments that relate to this dataset, each time matching the scale and cover.

For the experiments, we will simulate remapping an emission dataset using knowledge of traffic. First, it is necessary to have an emission data set, which we will generate by simulating an amount of traffic on the road network. The features contained in the dataset are polylines representing different segments of roads. The category associated with a segment s, c_s is used as an indicator for the amount of traffic: the bigger the category of road, the more traffic is assumed. An equal number of vehicles is assumed per km of a given category, so the length of the segment also is considered to generate the value. As a result, the simulated traffic for a segment s, t_s, is calculated for each segment as follows:

$$\forall s : t_s = length(s) \times (4 - c_s) \tag{11.1}$$

This yields a dataset which matches a realistic street layout, and contains a varying amount of traffic for each street section. To make our dataset more realistic, we also incorporated the knowledge on restricted and forbidden streets. Restricted streets should have less traffic, in our example 10% of the normal generated traffic for the

Fig. 11.1 Source data for generating the Warsaw dataset. Roads are indicated with different widths based on their category and coloured in orange or red if traffic if respectively restricted or forbidden. Parks and open spaces are shown in green, whereas large bodies of water are coloured blue

size is considered. Forbidden streets should have no traffic, so just 1% of its generated traffic for streets of this size is assumed. This knowledge is included by correcting the traffic for segment each s with a factor x_s:

$$\forall s : x_s = \begin{cases} 1 \text{ if segment } s \text{ has no restriction} \\ 0.1 \text{ if segment } s \text{ has restriction} = \textit{forbidden} \\ 0.01 \text{ if segment } s \text{ has restriction} = \textit{restricted} \end{cases} \quad (11.2)$$

The values for the traffic per street were multiplied with the correction factors to result in an adjusted traffic t'_s for each segment.

$$\forall s : t'_s = t'_s \times x_s \quad (11.3)$$

The resulting dataset is a feature based set containing road segments s; with each road segment a number t'_s is associated that will be used as an indication for traffic. This is shown on Fig. 11.2, with darker colours indicating higher values (more cars).

Next, it is necessary to simulate emissions e_s. In a perfect situation, the emission would have a one-one connection with the traffic, barring a constant factor or shift.

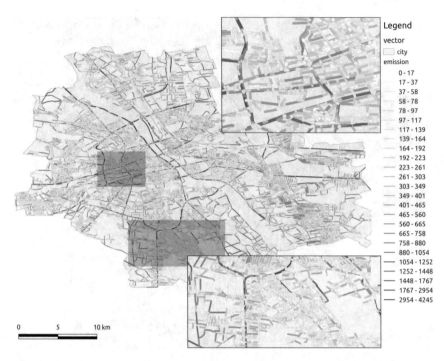

Fig. 11.2 Simulated traffic/emission based on road category and restrictions; darker colours indicate more traffic

The presented algorithm does not relate numerical values directly; the factor or shift are therefore irrelevant to the algorithm. As a result, we can also use the value t'_s to represent the amount of emission that perfectly matches the traffic.

$$\forall s : e_s = t'_s \tag{11.4}$$

This results in a set of road segments where each segment has a value representing emission data. For the experiments, it was opted that the goal is to remap simulated emission data using traffic data. This also makes it more intuitive to compare results when the emission data stays constant but the proxy data varies: all experiments should have a similar outcome. Now, it is necessary to revisit the traffic values: realistically, the emission measurements and traffic data will not have an ideal match, so a randomization of t'_s is introduced. Two aspects for randomizing the data were considered: randomization of the associated data (i.e. the value t'_s) and randomization of the spatial aspect (i.e. the geometry). To achieve the former, the traffic values of the road network were randomized using the formula below:

$$t^i_s = e_s + r_s \times e_s \times f_i \tag{11.5}$$

In this equation, e_s is the emission value for segment s, r_s is a random value in the range [0, 1] which is generated for segment s; each segment will have a differently generated value. The values f_i are used to determine how much the values t_s^i are allowed to deviate from e_s. Values between 0.1 and 1 were chosen, to allow for datasets where the randomized traffic differs from up to 10% to up to 100% of the original value.

This allows for randomizing values of segments and makes the values t_s^i correlate less with the values of e_s, the effect of this is lowered due to the fact that a grid cell tends to hold multiple segments. In addition, as most of the segments are quite short in length, it has a limited effect on the spatial distribution of the proxy data, keeping it too well centred around the generated emission data. To achieve a spatial randomization of the proxy data, buffers around the segments were created. These are considered around each segment and result in a polygon centered around it. The value t_s^i is then associated with these polygons. Different sizes of buffers allows us to further control how spread out the proxy data is. This is shown on Fig. 11.3, where the effect of four different sizes of buffers is shown.

Both the value randomization and the spatial randomization were combined, yielding a number of feature based datasets that can serve as the basis from which a proxy

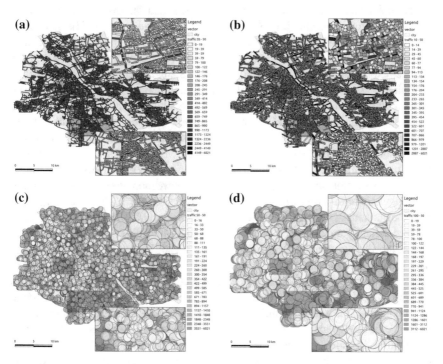

Fig. 11.3 Different sizes of the buffers, each with random factor $f_i = 0.5$. Due to the overlap of the buffers, many features are hidden

grid can be derived. In each of the experiments in the subsequent sections, the details on which generated data was used to derive the gridded datasets will be presented.

11.1.2 Algorithm Settings

The algorithm leaves many aspects where fine-tuning is possible. For the experiments in this chapter, the following choices were made

- generated training set
- minimum rank of a parameter $= 0.6$
- five linguistic terms for each variable, triangular linguistic fuzzy sets
- only proxy data used for parameters
- at most two variables

The training set is automatically generated for the test case. This is done using the algorithm described in Sect. 10.4. The main reason for this is that the Warsaw dataset is too small the supply both a training set and a data set while considering realistic grid-cell sizes. A secondary reason is that this is perhaps a more realistic situation for this type of datasets: the requirement that the overlap pattern has to match may limit the availability of suitable real world training sets. The generation of appropriate training data is therefor an important step to apply the approach in real world problems. Using the generated training set, the algorithm tries a number of parameters, and uses the approach from Sect. 8.3 to determine which parameters are suitable. The resulting value is not easy to interpret and was arbitrarily chosen. Experimental research revealed a realistic value to indicate similarity is 0.75, but for the experiments the threshold to select parameters was put lower than that at 0.6. This not only guaranteed that there would be parameters, but also provides a better illustration of what happens when the parameters are not ideal. This value is only used for determining which parameters to use, ongoing research aims at creating a weighted evaluation, to allow better parameters to have more impact. In the rulebase, five linguistic terms (*very low*, *low*, *medium*, *high* and *very high*) were considered over the domains of the parameters. This is an arbitrary choice, experimental research showed virtually no difference between 3, 5 and 7 linguistic terms. Each of the linguistic terms is represented by a triangular fuzzy set and the sets are spread out evenly over the domain. In none of the algorithms, parameters were defined on the input grid; only the proxy grid was used. The main reason for this is that the input grid is the lowest resolution grid in most tests and it provides therefor limited information. The main goal was to test the influence of the proxy data; some conclusions on the influence of proxy data are applicable on the input data as well. Lastly, at most two input variables were considered. Most of the parameters have some similarities (for the datasets used in the examples, all the highest ranked parameters considered overlap) and therefor will exhibit a similar behaviour. Adding more similar parameters increases the time needed to find a solution, without their being any real benefit. The developed method for constraint disaggregation, described in Chap. 7 is used in all experiments to disaggregate the results.

In experiments with multiple proxy data, the method set to take one parameter from each proxy data set. As many of the parameters that can be considered are quite similar, there would be a big risk that all the selected parameters relate to the same proxy data and that other proxy data would be ignored. By forcing the system to take the best parameter associated with each of the proxy data sets, the system has to take all proxy data into account. This may not be the best choice as some proxy data may not contribute much; more research in the equivalence of different parameters is necessary in order to allow the system to better judge if a dataset is helpful and if its use is preferred over the use of another parameter of an already considered dataset.

11.1.3 Datasets for Disaggregation

For the experiments, gridded datasets are needed. These datasets are derived by sampling the road network containing generated emission data and generated traffic data. Grids derived from the emission data form both the input set and the ideal output set, whereas grids derived from randomized traffic data will serve as proxy data.

Most test cases considered will define a grid of 20×20 cells over the envelope of the Warsaw area (each cell is $3.520 \, \text{km}^2$) as the output set. The main reason for this choice stems from the fact that this is a realistic size for the emission related research that spawned the development of the intelligent approach. In line with this, the input grid for a number of tests will be a grid containing 10×10 cells over the envelope of the Warsaw area, translating to grid cells that have an area of $14.018 \, \text{km}^2$. As such, the target grid partitions every cell of the input grid into four smaller cells. As the underlying distribution of the data is available, it is possible to generate the ideal solution of the 20×20 grid, allowing us to properly verify how close the developed algorithm can approximate the ideal solution. Some experiments will be run at higher resolutions, as these are more under influence of the spatial randomization.

These experiments will make use of a training set that is generated using the algorithm described on in Sect. 10.4. This is on one hand because our Warsaw dataset is not sufficiently large to have both a meaningful training set and test set, but also as this constitutes a more general application. For a disaggregation into a 20×20 for the Warsaw test data, one example of a generated training set is shown on Fig. 11.4.

The generated training set on Fig. 11.4 is one example; every experiment will generate its own training dataset in the same way: in each output cell, the number of points is randomly chosen from $\{0, 1, 4, 5, 9\}$ and each point has a random value (the range of which is determined by the range of the values in the input grid). By sampling this set of points using the geometry of input, output and proxy grids, a training set is obtained. For the current experiments, a single set of points was generated as it provided enough data.

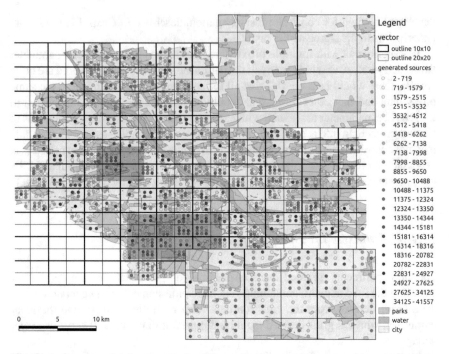

Fig. 11.4 One example of a generated trainingset for disaggregation into a 20 × 20 grid. The number of sample points in a cell is random, as is their value. The points are not positioned in parks or water

Important in judging the result is assessing their quality in comparison to an ideal solution. As explained in Chap. 8, it is difficult to determine a single value that can indicate the quality of a gridded spatial dataset: the goal of trying to approximate an underlying spatial distribution makes it necessary to not only consider the values of the cells but also where those values occur. The simple examples described in Chap. 8 elaborate on why neither Pearson correlation or least squared difference are suitable measures as they fail to take into account this spatial aspect. The method developed in Sect. 8.3 improves on this and is suitable to rank how well different grids match a given reference grid, but it is less suitable to judge a single grid as the value of the method is difficult to interpret. The similarity between two gridded approximations is a number in [0, 1], but the value is not easily to interpret on its own. Lacking suitable measures to properly assess the results, he main tool to judge the them is by analyzing the values and distributions individually in each of the experiments, using proper visualization tools and comparisons at grid cell level.

11.2 Disaggregation of Warsaw Test Data

This section contains different experiments to investigate the behaviour of the developed algorithm under different circumstances. Each of the subsections contains an experiment to investigate specific aspects of the proxy and/or input data.

11.2.1 Quality of Proxy Data

11.2.1.1 Connection Between Proxy Data and Input Data

The experiments in this section aim to show how the algorithm copes with deviations in the proxy data. For a first set of tests, a 10×10 grid is disaggregated into a 20×20. The proxy data was defined as a 23×23 grid at a $10°$ angle; the values of the traffic data were defined using a buffer of size 100 with values randomized using different factors f_i. The different cases are listed in Table 11.2.

For these tests, the common data are the input grid and the target grid. Both of these are shown on Fig. 11.5.

The cells are shaded according to their values. For both input and output, a Jenks natural break optimization [18, 46] is used to cluster the cells in 25 groups, with a darker shading as values increase. The choice of 25 groups is arbitrary, the reason for using the natural break optimization is to maximize the visual scale in the different

Table 11.2 Configurations for disaggregation experiments relating to deviations of the proxy data

Experiment	Input	Target	Proxy (traffic)	Factor f_i
1	10×10	20×20	23×23, $10°$	0.50
2	10×10	20×20	23×23, $10°$	0.75
3	10×10	20×20	23×23, $10°$	1.00

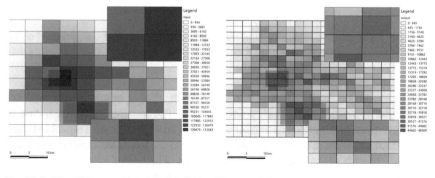

Fig. 11.5 10×10 input grid and the ideal 20×20 output grid

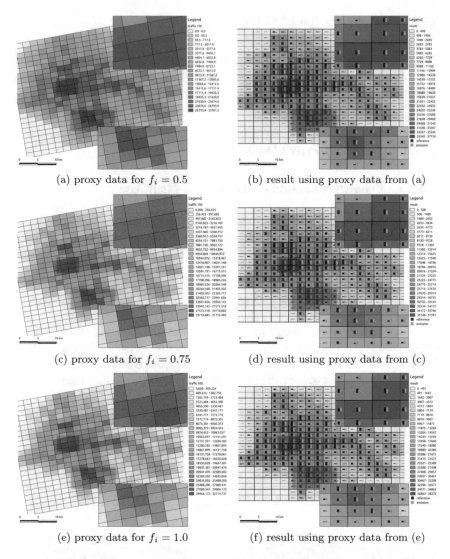

(a) proxy data for $f_i = 0.5$ (b) result using proxy data from (a)

(c) proxy data for $f_i = 0.75$ (d) result using proxy data from (c)

(e) proxy data for $f_i = 1.0$ (f) result using proxy data from (e)

Fig. 11.6 Proxy grid and results for disaggregation with proxy data that was generated on the randomized buffered road network using a buffer of size 100

figures. A consequence is that due to this choice, shades in one grid should not directly be compared with those in another grid. The training set for each case in this experiment is generated as explained in Sect. 10.4. The varying aspect between these examples is the proxy data, which is shown to the left of the matching result in Fig. 11.6. In each of the results, the ideal solution is compared against the obtained result using a bar chart in each grid cell: the left bar indicates the reference (the value of the cell in the ideal output grid) whereas the right bar indicates the result.

While it is tempting to immediately compare the bar-charts, the reference solution will be very hard to obtain. Of interest is how well the algorithm managed to achieve the ideal distribution in each of the input grid cells. Questions regarding the resemblance of the disaggregation pattern (is the lowest/highest cell in the disaggregation of an input the same in the result and in the reference) are of importance to judge the result.

At first sight, there is little between these three results. However, some differences appear in the center of the lower enlarged section. The solution in Fig. 11.6b, the example that uses $f_i = 0.5$, shows a more divided distribution in the lower center input cell: the algorithm estimates higher values on the left side, deviating more from the ideal solution. By contrast, the result in Fig. 11.6f, using $f_i = 1$ shows a more even distribution in this location. This is explained by the pattern in the proxy data: the proxy data in Fig. 11.6a, with $f_i = 0.5$ has higher values that result in the algorithm mapping the data in that location. It is merely a coincidence that the proxy data generated while allowing higher random factors (Fig. 11.6e, with $f_i = 1$) resembles the ideal solution better. In general, we observe that the solution in Fig. 11.6f has a less pronounced disaggregation, and values are assigned more equally to all partitions of a cell. This is noticeable in the top area that is enlarged, where four input cells and their resulting partitions are visible. From the center of the main image, there appears to be a horizontal pattern stretching to the right (just below the bottom right corner of the top enlarged area). In Fig. 11.6b, the disaggregation is such that it favors higher values in the lower half of the cell, matching to some extent the ideal values. In Fig. 11.6f, this pattern is still visible, but it is less pronounced; while the disaggregation still assigns higher values to the lower half, the difference between lower half and upper half is smaller. The explanation for this is of course that the higher randomization factor causes the proxy data to be less matched with the input data, and as such it does not provide as good information to help perform a spatial disaggregation.

11.2.1.2 Impact of Proxy Resolution

The set of tests in this experiment investigate the impact of the resolution of the proxy data. It stands to reason that higher resolution proxy data will yield better results, but it is interesting to investigate if low resolution proxy data can still help somewhat, and how the system copes with proxy data that has too low a resolution. For this purpose, the input and output grid are selected as before (respectively 10×10 and 20×20, Fig. 11.5), with the proxy data based on the buffered randomization with a buffer size 100 and factor $f_i = 0.5$ (Fig. 11.3d). The same underlying randomized proxy data was used for all these experiments (Table 11.3).

The first set of tests, displayed in Fig. 11.7 uses low-resolution proxy data, ranging from 10×10 (the same size as the input grid) to 23×23 (slightly more than the 20×20 output grid).

Table 11.3 Configurations for disaggregation experiments relating to different resolutions of the proxy data

Experiment	Input	Target	Proxy (traffic)
1	10×10	20×20	$10 \times 10, 10°$
2	10×10	20×20	$16 \times 16, 10°$
3	10×10	20×20	$23 \times 23, 10°$
4	10×10	20×20	$35 \times 35, 10°$
5	10×10	20×20	$50 \times 50, 10°$

In many of the input cells, particularly in Fig. 11.7b, the disaggregation over the different partitions is fairly limited: the yellow bars in most of the output cells that partition the same input cell are quite similar in height (the shade of the cells is a bit misleading in some cases due to the natural Jenks clustering). A notable exception is near the bottom center, just below the bottom left corner of the lower enlarged area, where the right cells have higher values than the left cells. The reason for this is obvious when looking at the proxy data: at this location, the proxy data not only divides the input cell but the cells involved also have differing values. This is reflected in the disaggregation. A similar situation is visible in the top left area.

The results shown in Fig. 11.7e, show a similar pattern: some input cells are disaggregated in almost equal values, whereas others seem to have been disaggregated more accurately. Particularly in the center the proxy data does not help much: due to its rotated position around the center, this is the area where the geometry of the cells of the proxy data most closely match the input data, which is the case when parameters that use overlap cannot help the disaggregation.

Tests with higher resolution proxy data, e.g. the experiments on Fig. 11.8 show disaggregations that better match the ideal result in more locations. The number of input cells that are disaggregated more accurately increases as the resolution increases. This is particularly visible in the top left of the top enlarged area and in the lower center input cell depicted in the lower enlarged area.

The contribution that low resolution proxy data can make on a spatial disaggregation with the presented method greatly depends on the intersection pattern of input data and proxy data: in places where the proxy grid is such that it does not divide an input cell, then the data cannot help with the underlying resolution. There may be parameters conceivable that would still manage to disaggregate better in this situation, but more research in new parameters is needed as none of the current set of considered parameters was capable of this. Higher resolution proxy data, when compared against the input grid, divides the latter in more locations and as such provides more information.

In addition, these experiments show what happens when proxy data cannot provide much information: the data of an input cell is distributed almost equally over its output cells. Visually, this can deviate a lot from the ideal solution, but barring any other information it is the most natural choice for disaggregation.

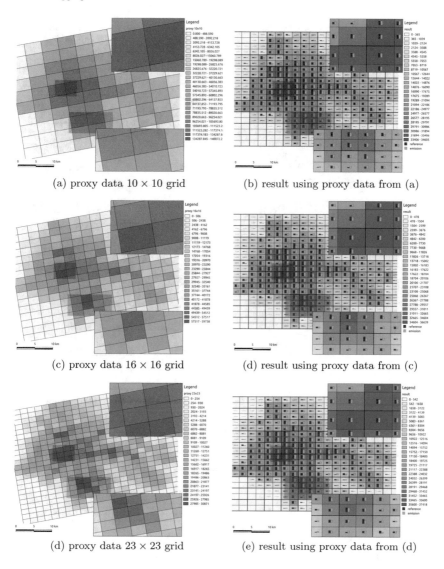

(a) proxy data 10 × 10 grid   (b) result using proxy data from (a)

(c) proxy data 16 × 16 grid   (d) result using proxy data from (c)

(d) proxy data 23 × 23 grid   (e) result using proxy data from (d)

Fig. 11.7 Proxy grids and matching results for disaggregation with proxy data of different resolutions

11.2.1.3 Impact of Proxy Angle

In this set of tests, the resolution of all grids involved is constant, but the angle of the proxy grid changes. The input and output are the same as in Sect. 11.2.1.2 and shown on Fig. 11.5; the underlying randomized proxy data is the same for all four tests. Details of the four experiments are listed in Table 11.4.

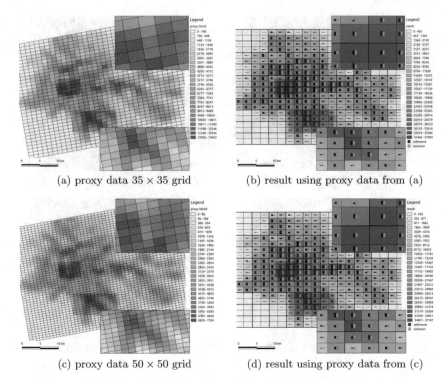

(a) proxy data 35 × 35 grid   (b) result using proxy data from (a)

(c) proxy data 50 × 50 grid   (d) result using proxy data from (c)

Fig. 11.8 Proxy grid and results for disaggregation with proxy data of different resolutions

Table 11.4 Configurations for disaggregation experiments relating to the orientation of the proxy data

Experiment	Input	Target	Proxy (traffic)
1	10 × 10	20 × 20	31 × 31, 10°
2	10 × 10	20 × 20	31 × 31, 20°
3	10 × 10	20 × 20	31 × 31, 30°
4	10 × 10	20 × 20	31 × 31, 45°

The results are shown on Figs. 11.9. The proxy data is overlayed with the input grid (bold lines) and output grid (dashed lines) in order to better show they relate. Bear in mind that the shades of the different proxy grids cannot be compared, but within a grid a darker shade always represent a higher value in a grid.

As in the previous experiment, the difference is in the details. The input cell located in the top left of the lower enlarged area is a good place to start. The first and third experiment assign the top right output cell of this input cell the highest of the four values; the bottom left cell receives the lowest value while the other two cells

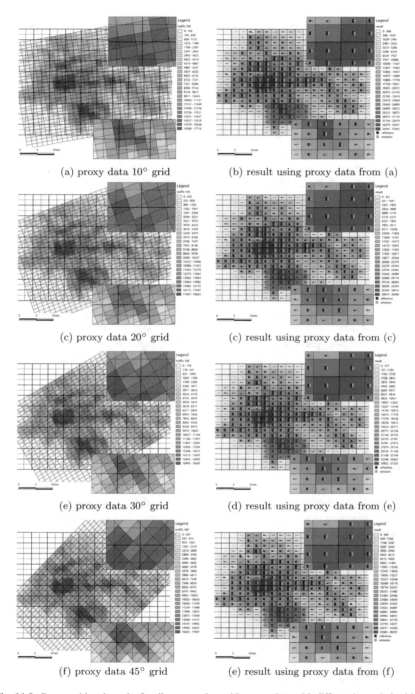

(a) proxy data 10° grid

(b) result using proxy data from (a)

(c) proxy data 20° grid

(c) result using proxy data from (c)

(e) proxy data 30° grid

(d) result using proxy data from (e)

(f) proxy data 45° grid

(e) result using proxy data from (f)

Fig. 11.9 Proxy grid and results for disaggregation with proxy data with differently angled grids. In the proxy data (**a**), (**c**), (**e**) and (**f**) the input grid and output grid are shown respectively using bold and dotted lines

are similar. The second and fourth test also assign the lowest value to the bottom left cell, but treat the right two cells more or less equal. When considering the grids, the underlying spatial distribution is not known, but the different proxy grids in Fig. 11.9 make it appear as if there is a higher value in the right and top right area of this input cell. Depending on which proxy grid is used, this is either more towards the top right, or towards the entire right, which is reflected in the result.

A similar situation is observed in the bottom right input cell of the lower enlarged area. All four tests assign its top right output cell the lowest value. The first test however assigns the bottom left output cell the highest value, the second test considers the three bottom left output cells more or less equal, while the last two tests put most weight on the left cells. The bottom center input cell in this enlarged area shows similar results. This again matches with the patterns observed in the proxy data. As in the previous experiment, the way the proxy data intersects the input grid has an influence.

In a realistic situation, the underlying distribution is not known, and only of the proxy grids would be available. The issue of not knowing the underlying distribution of a gridded dataset arises here in the proxy data, as this experiment clearly shows. The current parameters calculated on the proxy data implicitly assume a uniform distribution of the data within each cell of the proxy grid; this assumption causes the behaviour observed in this experiment. The main issue is way the proxy data and input data overlap, particularly the overlaps where input data is partitioned. Proxy cells that partition input cells affect how the data in the output will be distributed. This also means that the effect decreases as the resolution of the proxy data increases. Later experiments will also show that the use of multiple proxy also decreases this effect.

11.2.2   Disaggregation to Higher Resolutions

In the previous experiments, the impact of changes to the proxy data was considered. Here, the attention goes to the limits of the disaggregation: is it possible to disaggregate each input cell into more than four output cells? For these experiments, both the input and proxy data will be fixed: the former is again a 10×10 grid (Fig. 11.5a) whereas the latter is a 31×31 grid (Fig. 11.9a) (Table 11.5).

In the result grids, shown in Fig. 11.10, the bar charts are omitted as they would be too small but also to better show the patterns that appear in the calculated solution.

Slightly showing in the 30×30 grid, but more prominently visible as the target resolution increases, are patterns of cells with similar value. This actually means that in those areas, the underlying data is of lower resolution than that at which it is represented. While the output grid has the desired resolution, this pattern reveals that the proxy data is not sufficient enough to remap the data onto this desired resolution. One could say that the result of the regridding is a grid of lower underlying resolution than the output grid specified. This mostly occurs in places where two or more output cells are covered by the same proxy cell as the parameters used mainly consider

Table 11.5 Configurations for disaggregation experiments relating to the resolution of the proxy data

Experiment	Input	Target	Proxy (traffic)
1	10×10	20×20	$31 \times 31, 10°$
2	10×10	30×30	$31 \times 31, 10°$
3	10×10	40×40	$31 \times 31, 10°$
4	10×10	50×50	$31 \times 31, 10°$

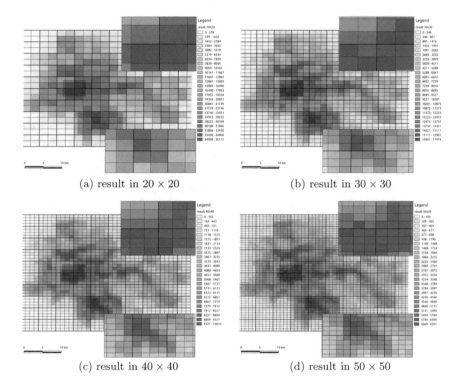

(a) result in 20×20

(b) result in 30×30

(c) result in 40×40

(d) result in 50×50

Fig. 11.10 Results for disaggregation in different resolutions

overlap or overlap over a wider region. If there are not many differences in the calculated parameter values, then these output cells will have very similar values for there variables in the rulebase system. The result is that the disaggregation in these areas yields equal values, similar to what was observed in Sect. 11.2.1.2 when the resolution of the proxy data is too low. These experiments highlight that the there are limits to the resolution to which a disaggregation can be performed, and this resolution is dependent on the proxy data. By using proxy data of higher resolution or quality and even by using multiple datasets of proxy data, the output resolution can be increased.

11.2.3 Multiple Proxy Data

The previous experiments show that the proxy data often manages to steer the disaggregation. Of particular interest is the possibility to use more than one set of proxy data. This can help to alleviate some the issues relating to the overlap patterns that were observed in the experiments in Sect. 11.2.1.

Three experiments that make use of multiple proxy data are presented in this section. The details of these experiments are listed in Table 11.6. All these experiments disaggregate a 10×10 grid into a 20×20 grid, both were shown earlier on Fig. 11.5.

The first experiment, of which both the proxy grids considered and the result are shown on Fig. 11.11, aims at illustrating the use of various different proxy grids: the proxy grids have different resolutions and/or orientations.

The results are better than those that used a single 20×20 rotated proxy grid, as tested in the experiment in Sect. 11.2.1.2. The algorithm manages to find the low value cells near the middle of the main map, and also quite accurately matches the diagonal line near the right of the main map. The enlarged sections confirm this, with one exception in the bottom-center of the lower enlarged area: here, the algorithm did detect that the top two output cells should have greater values than the bottom two, but the bottom left value is not as low as it should be (while the top right value is slightly too high). The reason for this is in the proxy data: all three proxy grids show higher values in the top half of the input cell, but do not show that much difference between the left and right halves. The visual pattern of a vertical line meeting a horizontal line appears therefor slightly blurred. This behaviour is a consequence of the limited representation in the proxy grid: within the proxy grid, there is knowledge regarding the underlying spatial distribution thus data is spread out over its cells.

Table 11.6 Configurations for disaggregation experiments that involve multiple proxy grids

Experiment	Input	Target	Proxy (traffic)
1	10×10	20×20	20×20, $10.0°$
			25×25, $25.0°$
			30×30, $45.0°$
2	10×10	20×20	10×10, $10.0°$
			10×10, $20.0°$
			10×10, $30.0°$
			10×10, $40.0°$
			10×10, $50.0°$
2	10×10	20×20	20×20, $10.0°$
			20×20, $20.0°$
			20×20, $30.0°$
			20×20, $40.0°$
			20×20, $50.0°$

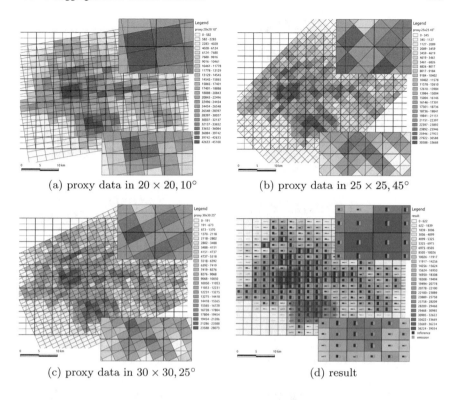

(a) proxy data in $20 \times 20, 10°$ (b) proxy data in $25 \times 25, 45°$

(c) proxy data in $30 \times 30, 25°$ (d) result

Fig. 11.11 Proxy data (**a**), (**b**) and (**c**), overlayed with the input grid (bold lines) and output grid (dotted lines) and the result of disaggregation using this proxy data (**d**). The bar charts in (**d**) compare the reference (ideal) solution with the calculated emission

The second experiment uses many low resolution proxy grids: five proxy grids of 10×10 at different angles are used to disaggregate a 10×10 grid into a 20×20 grid. While the experiment in Sect. 11.2.1.2 showed that the contribution of a low resolution proxy grid is limited, this experiment aims to verify if combining multiple low proxy grids with different overlap patterns will contribute. In this case the proxy grids do not have a higher resolution than the input grid, but due to their different orientation they each provide different information on the underlying spatial distribution. The different proxy grids are shown on Fig. 11.12a–e; the result is on Fig. 11.12f.

The results can be compared with those in Sect. 11.2.1.2, particularly with the low resolution tests, shown on Fig. 11.7: the proxy grids in this experiments have the same resolution as the proxy grid in Fig. 11.7a. The result in Fig. 11.12 is better than that in Fig. 11.7b.

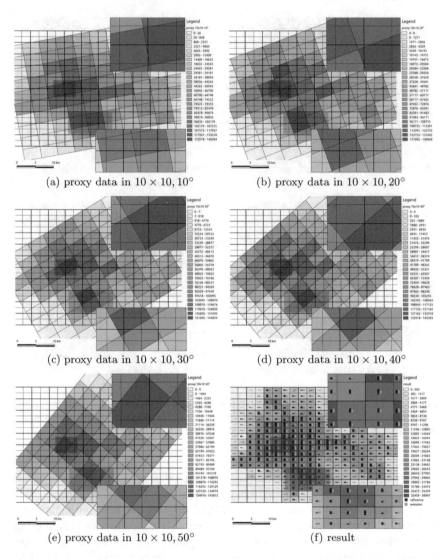

Fig. 11.12 Second experiment for disaggregation using multiple proxy grids. Different low resolution proxy grids (**a**)–(**e**) are used to perform the regridding; the result is shown in (**f**)

For the input cells that are highlighted in the top enlarged section of the map, the disaggregation is better than when only using a single proxy grid. Here, as it is in the center of the map, all grids overlap and furthermore there are many different overlap patterns, all providing different but helpful information.

There are some input cells where a disaggregation is almost not possible. This is for example the case in the third input grid in the top row: all output cells have nearly-equal values. In this location, the only grid that provides information is the proxy grid (a). As the other grids do not overlap, they do not provide helpful information which in the current implementation is not ignored. As such, the helpful information provided by one grid is offset by the lack of information from the others causing the result for this input cell to actually be worse than that shown in Fig. 11.7b. A similar observation can be made in the bottom center input cell of the second enlarged area: the system wrongly estimates the value of the top left output cell as the highest cell, wheres the calculation using just one proxy grid (Fig. 11.7b) does not show this pattern. The main culprit are the proxy grids in Fig. 11.12b and d. The algorithm follows all patterns, so it is a consequence of the proxy grids and how they approximate the underlying distribution.

This experiment shows that low resolution proxy data can contribute but has its limitations: different intersection patterns may result in sub-optimal outcomes in specific areas of the grid. As mentioned in Sect. 11.1, the current implementation is forced to use all the proxy grids it is given (mainly to illustrate such side-effects), but this example shows that it should be possible to ignore data from other grids if it proves to have a negative effect. As such, the weighted evaluation of the rulebase, which was hinted at in Sect. 11.1, should not have global weights assigned for the variables, but these weights should also be allowed to vary for different locations. Allowing parameters to be defined on the input grid is similar to using multiple proxy grids. The input grid will cannot provide additional information on its own underlying distribution, so the effect will mainly be that its inclusion will tend to smoothen the results. This can be beneficial as it prevents the algorithm of following the proxy grid(s) too strictly.

The third experiment employs many higher resolution proxy grids; this result can be compared against those in Fig. 11.8 and is shown on Fig. 11.13. The outcome of the algorithm is quite good, but in some places worse than the result obtained using a single high resolution proxy grid. This is particularly the case for those input cells for which not all proxy grids contribute (e.g. near the top of the map), but also for some of the other input cells. The former issue was mentioned before: the lack of information from a proxy grid will equalize the results of the disaggregation. The latter also makes sense: more data is present, but the data is somewhat contradictory due to the fact that each grid is a different spatial approximation of an underlying distribution. The consequence is that the disaggregation using multiple proxy grids tends to be less pronounced and more conservative. This is a good thing for stability: a small change to one proxy grid will not cause huge changes in the end result, and while it limits the possibility of disaggregating into very extreme values within an input cell, it also prevents making big errors in the disaggregation.

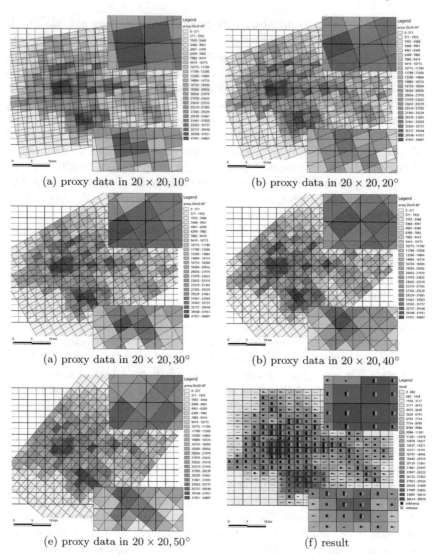

(a) proxy data in $20 \times 20, 10°$ (b) proxy data in $20 \times 20, 20°$

(a) proxy data in $20 \times 20, 30°$ (b) proxy data in $20 \times 20, 40°$

(e) proxy data in $20 \times 20, 50°$ (f) result

Fig. 11.13 Proxy data and outcome for the third experiment with multiple proxy grids. The proxy data has all are defined on a 20×20 grid, but the cases differ in angle: **a** $10°$, **b** $20°$, **c** $30°$, **d** $40°$, **e** $50°$. The result is shown on (**f**)

11.3   Regridding of Warsaw Test Data

The algorithm is not only capable of performing spatial disaggregation, but also of regridding, by using first deriving a new grid that partitions both input and output grids (Sect. 2.2). A few examples of regridding are shown, based on the same generated

dataset as before. The main purpose is to show off the possibilities. Due to the partial
overlap between input cells and output cells, there are more degrees of freedom in
mapping the data and judging the results requires a look at the overall picture rather
than at individual output cells.

11.3.1   Angle of the Input Grids

An overview of the tests performed in this section is in Table 11.7. The four considered
tests use the same proxy data and the same target grid, but the input grid varies: it
has the same resolution in all cases, but is angled differently.

The common data for these experiments are the target grid and the proxy grid. Both
are shown on Fig. 11.14, with the output grid shaded to illustrate the ideal solution.
The proxy grid uses factor $f_i = 0.5$ for randomizing the data values, combined with
a buffer size of 100.

One necessary step in the regridding process is the creation of a segment grid
(Sect. 2.2). For the first two experiments, the segment grid is shown on Fig. 11.15.
This segment grid is generated from the intersection of both input and output grid and

Table 11.7 Configurations for regridding experiments using an input grid with incompatible grid
layout and varying angles

Experiment	Input	Target	Proxy (traffic)
1	$13 \times 13, 0.0°$	20×20	$31 \times 31, 10°$
2	$13 \times 13, 10.0°$	30×30	$31 \times 31, 10°$
3	$13 \times 13, 25.0°$	40×40	$31 \times 31, 10°$
4	$13 \times 13, 45.0°$	50×50	$31 \times 31, 10°$

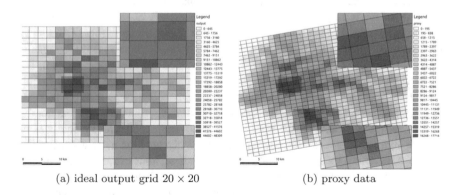

(a) ideal output grid 20×20 (b) proxy data

Fig. 11.14 Target/ideal output grid (**a**) and the proxy data (**b**) for the experiments that relate to
regridding a rotated input grid

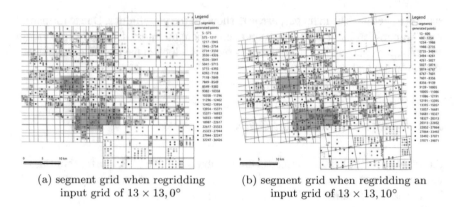

(a) segment grid when regridding   (b) segment grid when regridding an
input grid of $13 \times 13, 0°$ input grid of $13 \times 13, 10°$

Fig. 11.15 Example of the segments grid used for regridding. The data points for the generated training sets are also indicated

used as a initial disaggregation target. As in the previous experiments, the training set is generated for each application and Fig. 11.15 also shows the generated point set used to create the training sets for the examples.

Both the input grids and output grids of the different tests are shown on Fig. 11.16. The shading is somewhat misleading as a comparison between the different results, so the focus should be on the bar charts. In addition, the rotated input grids do not cover the same area as the output grids, so near the top left and bottom right there are cells that do not have a value in the calculated output, as there was no matching value in the input. Consequently, as there was no input data, the errors here are normal and for comparison we should look at areas for which all the input grids have data.

The results are in line with the previous observations. First, consider the top enlarged section. All but the last result in Fig. 11.16h slightly overestimate the value of the cell in the center-bottom right. The results in Fig. 11.16d and f underestimate the value of the top right, whereas the other two slightly overestimate it. We also see the same pattern in the bottom enlarged section: the results in Fig. 11.16d and f show a similar pattern for the cells in the center, whereas those cells in the results of Fig. 11.16b and h also are similar. The explanation for this similarity is in the intersection patterns and the implicit assumption that the data in a grid cell in uniformly distributed. In the regridding problem, data associated with an input cell that overlaps two output cells can be mapped outside of the output cell (this was not the case in disaggregation, as there were no partial overlaps). The proxy data was of relative high resolution, so this effect is limited. When we look more general at the solutions, there are many places where the pattern of values in the output grid is matched, showing that the algorithm is capable of correctly assessing values in such areas. This is particularly noticeable in areas where there are big differences, e.g. near the edges of the city.

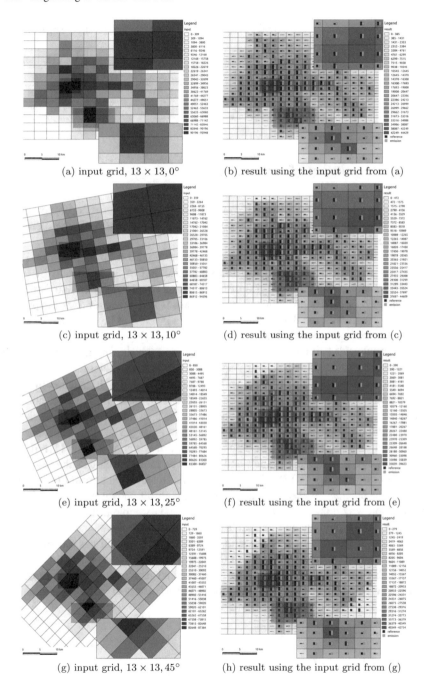

(a) input grid, $13 \times 13, 0°$ (b) result using the input grid from (a)

(c) input grid, $13 \times 13, 10°$ (d) result using the input grid from (c)

(e) input grid, $13 \times 13, 25°$ (f) result using the input grid from (e)

(g) input grid, $13 \times 13, 45°$ (h) result using the input grid from (g)

Fig. 11.16 Results for regridding rotated grids

Table 11.8 Configurations for regridding experiments using an input grid with incompatible grid layout and varying angles

Experiment	Input	Angle	Target	Proxy (traffic)
1	13 × 13	25.0	20 × 20	31 × 31, 10°
2	18 × 18	25.0	20 × 20	31 × 31, 10°
3	29 × 29	25.0	20 × 20	31 × 31, 10°

11.3.2   Resolution of the Input Grid

The tests in this section are aimed at verifying the regridding process of input grids of different resolution. Three tests are considered, with the details of the experiment listed in Table 11.8.

In this experiment, the regridding is performed to a 20 × 20 output grid. Three scenarios are considered: test 1 performs a regridding of an input grid of lower resolution, test 2 uses an input grid of more or less equal resolution and test 3 has an input grid of higher resolution. The latter case is also interesting, as the goal of regridding is to alter the data so that it matches a given grid layout; there can be a need for lowering the resolution.

The input grids and their regridded solutions are both shown on Fig. 11.17, with bar charts comparing the ideal values against the calculated values.

The result of regridding the lower resolution input (Fig. 11.17b) matches the ideal output quite well. In the top left of the top enlarged area, the regridding does not fully match the ideal value, but the relative values is this area match the ideal. A similar observation can made in the top left of the bottom enlarged section. The same test performed on a higher resolution input, the result of which is on Fig. 11.17d, puts the values closer to the ideal values and this is even more the case as the resolution increases further (Fig. 11.17f). The outline of the diagonal pattern in the center of the bottom enlarged section is least clear on the result shown in Fig. 11.17d; the explanation why it is clearer on the lower resolution example is in the combination of proxy data, input data and intersection patterns: the input data intersects in such a way that an interpretation of the grid as uniform in each cell has the side effect that concentrates the data more. Both the lower and the higher resolution happen to have distinctive cells that allow for the diagonal pattern to be more separated from the vertical pattern. This is similar to what was observed in the experiments in Sect. 11.2.1.

11.3.3   Multiple Proxy Data

This experiment is aimed at demonstrating some of the capabilities and the limits. Two experiments perform higher resolution regridding using multiple proxy grids.

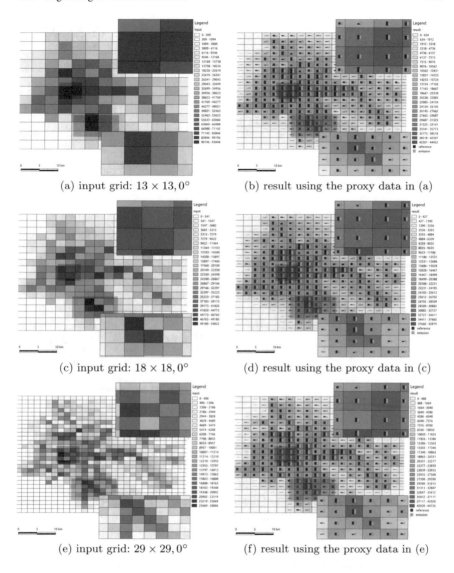

(a) input grid: $13 \times 13, 0°$ (b) result using the proxy data in (a)

(c) input grid: $18 \times 18, 0°$ (d) result using the proxy data in (c)

(e) input grid: $29 \times 29, 0°$ (f) result using the proxy data in (e)

Fig. 11.17 Different resolutions of input grids and the associated results

The higher resolutions makes the spatial randomization more effective (in these experiments, a buffer of 100 and a factor $f_i = 0.5$ are used), resulting in proxy data that has a worse connection and thus more difficult regridding. This makes for the fact that the proxy grids are of lower quality for the problem considered and this will be reflected in the results. The parameters for both tests related to regridding using multiple proxy grids are shown on Table 11.9.

Table 11.9 Configurations for disaggregation experiments that involve multiple proxy grids

Experiment	Input	Target	Proxy (traffic)
1	33 × 37, 10.0°	40 × 40	20 × 20, 10.0°
			25 × 25, 25.0°
			30 × 30, 45.0°
2	33 × 37, 10.0°	40 × 40	20 × 20, 10.0°
			20 × 20, 20.0°
			20 × 20, 30.0°
			20 × 20, 40.0°
			20 × 20, 50.0°

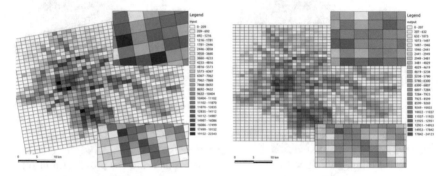

Fig. 11.18 33 × 37 input grid, angled 10° and the ideal 40 × 40 output grid

The first test uses multiple grids of differing sizes and orientation; the second test uses five grids of differing orientation but with the same resolution as the output grid. This test is similar in concept to the test in Sect. 11.2.3. The input and output grids for these experiments are shown on Fig. 11.18.

The proxy grids and outcome of the first test are shown on Fig. 11.19. The overall values are quite close, with two interesting remarks. First, in the bottom enlarged section, the vertical pattern is recognized. The ideal values (the red bars) but it mainly at the third cell from the left, whereas the algorithm positions the main values on the cell next to that. Analysing the data more closely reveals that the underlying distribution is an emission source (road, the network is shown on Fig. 11.1) that is on the right of the third cell, right next to the fourth cell. While the algorithm has it correct for any of the cells of that vertical pattern (as shown on the main map, e.g. below the enlarged section), it gets it wrong in some locations. This is caused by the intersection patterns and the gridded approximation: the angled 33 × 37 input grid has this vertical pattern in one column as can be seen on Fig. 11.18. In addition, two of the three proxy grids used (Fig. 11.19a and c) exhibit the same pattern, as they are angled similarly. This makes the system react as if there is an angled underlying

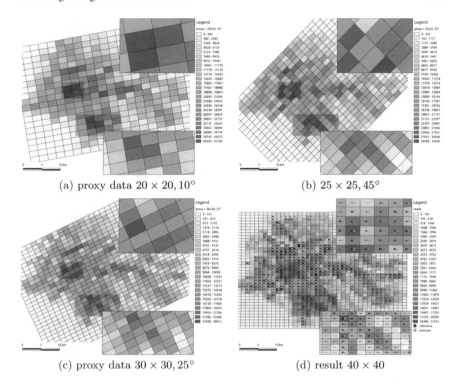

(a) proxy data 20 × 20, 10°   (b) 25 × 25, 45°

(c) proxy data 30 × 30, 25°   (d) result 40 × 40

Fig. 11.19 Different proxy grids (**a**), (**b**) and (**c**); and the result (**d**)

pattern rather than a vertical. It should be noted that it does this only for those 3 cells in the enlarged section and one cell below it. Further down the map, the algorithm correctly assess the values, as it is also better reflected in the proxy data.

The second test, for which the proxy data and result are shown on Fig. 11.20 shows similar patterns. The proxy data is of lower resolution than in the previous test, but more data is used. As the data is all mainly angled in the same direction, the erroneous conclusion that the vertical pattern is for some part in the cell next to the cell where it ought to be is also occurring here.

While the location of the data is not far off from the real location, the gridded approximation and its interpretation gives the impression it is further off. In addition, the algorithm in these experiments was forced to use each of the proxy grids, even if some of them may have worse parameters. This was done to prevent the algorithm from selecting only parameters connected to a single proxy grid and thus ignoring the information from other proxy data. This is an important decision. A grid can have its parameters considered of lower quality e.g. because the grid has a lower resolution, but at the same time it can provide valuable information that is not contained in the other proxy grids (not really the case in these examples, but in general). At the

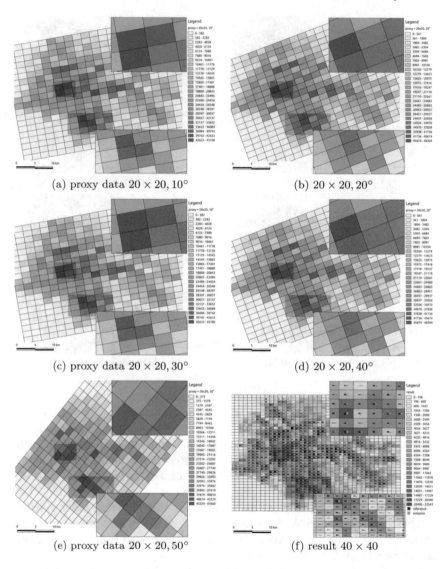

Fig. 11.20 Different proxy grids (**a**)–(**e**); and the result (**f**)

same time however, the lower quality parameters can negatively affect some places in the grid. Ideally, the variables derived from the parameters should be weighted in order to prevent the negative influence in some areas, and allowing for this is part of ongoing research.

11.4   Concluding Remarks

The experiments highlight both the strengths and weaknesses of the developed method. Strong points are that the type of proxy data is very wide: grids of any resolution can be used. In addition, they can be used to approximate feature-based data, so any dataset can potentially be used as proxy data. The current implementation already has partial support for proxy data to be incorporated without resorting to a conversion to a raster (the proxy data water and parks in the example were not converted to grids), and ongoing research aims at increasing the ways in which feature based data can be directly used.

   The suitability of a dataset for use as proxy data is determined automatically beforehand (Sect. 5.3) and if a dataset is not appropriate, it is not used. The method tries to use the knowledge of the spatial distribution of all available data (input and proxy data) to determine weights for a spatial disaggregation. Regridding problems are first transformed to a problem of spatial disaggregation. The knowledge of the different data is not weighted, which is seen on the experiments that involve multiple proxy data. As such, it is important not to allow the system to use bad quality parameters, and it makes selecting the parameters Sect. 5.3 important. The system works better with one good proxy data than with a good one and a bad one. A better solution would be to weigh the impact of the different parameters, as this would allow less suitable parameters to have less impact. The typical Mamdani rulebase systems however do not allow for weights in different aspects of the rules; incorporating the suitability of parameters in the evaluation is therefore not a straight forward task. In addition, this weighing should be defined globally for the rules: in some locations, some grids may be more suitable than others, as highlighted by the experiments in Sect. 11.2.3. Both the selection of suitable parameters as well as the possibilities for weighing the impact of the different parameters and priorities of future research. One of the aspects under consideration is to use quantifiers in the rules, to allow predicates such as *if most of the parameters are high then...*, as this allows for a form of consensus building using different proxy data and may better cope with proxy data that contains missing data (e.g. in the case of the rotated grids).

   Little time was spent in optimizing the implementation at this point. The main reason for this is that the novel methodology is still very much under development and optimized code tends to be more difficult to maintain. Some small optimizations have been introduced, mainly to e.g. to limit recalculation of values. More optimizations are possible; the most important aspect of the algorithm is that it is linear in the number of output cells considered, and that the algorithm is suitable for parallel processing: every output cell can be calculated independently of other output cells. To achieve linearity, it is necessary to make sure that the minimum and maximum possible values of parameters are cached when they are global (Sect. 5.2.1).

   The main focus was on spatial disaggregation and regridding, however the way the algorithm deals with proxy data, and in particular the way the disaggregation result is determined indirectly through the computation of weights allows the method to be extended for other applications. In particular, other applications are considered for data fusion and for the identification of locations that meet certain criteria.

Chapter 12
Conclusion

Abstract This final chapter summarizes the results of the research, and provides an overview of the different advancements in the different fields. In addition, directions for future research are listed; this not only concerns research in spatial data processing but also in related fields that were necessary to develop the current algorithm. Lastly, new application fields for the methodology are presented: the concept is quite universal and there are other applications that could benefit from a similar approach.

12.1 Novel Aspects Presented in This Work

The research presented in this book summarizes the efforts of the author to employ artificial intelligence to improve spatial data and spatial data processing. The map overlay problem for gridded data, which occurs in spatial disaggregation and regridding (Chap. 2). The origin of the problem stems from the fundamentals of gridded data, approximating the two dimensional space with a discrete model. The current methods for regridding work based on assumptions regarding the underlying spatial data distribution. This passes by on the fact that there is a vast amount of data on various topics available, and that past research has found correlations between this data. As such, the presented method tries to involve other datasets, proxy data, to estimate the underlying distribution, rather than basing this on pure assumptions. The current concept (Sect. 3) is aimed at trying to find correlations between other datasets by trying different parameters that are calculated based on the geometries and values of the grids. To make this work, a number of new developments in the field of rulebase systems related topics was required; this was elaborated on in Part II.

Many of the developments that were necessary in the context of processing the spatial data can be applied in other fields as well. The first novel aspect is a novel way in which rulebases can be constructed and applied using dynamic linguistic terms. In traditional rulebase construction, the domains of the variables in the rules and linguistic terms on these domains are determined in advance and kept fixed over both construction and application. The methodology explained in Chap. 6 generalizes this by connecting the domain of a variable to the data that is considered. This was a necessary step to allow for the fact that data in spatial datasets can be heterogeneous:

J. Verstraete, *Artificial Intelligent Methods for Handling Spatial Data*, Studies in Fuzziness and Soft Computing 370, https://doi.org/10.1007/978-3-030-00238-1_12

properties can be more present in one location than in another and using globally defined domains limits the expressibility. This change may however be applicable in for other problems. For the spatial problem, different possibilities for determining an optimal range were presented. While the exact calculations cannot be generalized as the determination of specific ranges to be used depends on the problem at hand, the two presented approaches (local range and estimated range, Sects. 5.2.2 and 5.2.3) can be applied for other cases. The second novel development also ties in with fuzzy research and is related to defuzzification. A method was developed (Chap. 7) to defuzzify multiple fuzzy sets that share a common constraint—in our case the sum of the defuzzified values is a known, crisp value. This approach maximizes the minimum of the membership grades of the defuzzified values, thus selecting the best fitting defuzzification for the different sets. While the specific problem is less common, the developed methodology is independent of the application and more broadly applicable.

In Chap. 8, a technique for comparing the quality with which grids resemble other grids is presented. This was necessary as many currently used methods ignore the spatial aspects and merely consider the grid as a set of value. Specific to this approach is that it estimates possible unknown underlying distributions and favours grids that seemingly approximate the same underlying distributions. By taking this into account, the developed method is capable of better assessing the quality of gridded approximations. The method returns a number in [0, 1] to indicate how well the grid resembles a reference grid. While the number in itself is difficult to interpret, it is useful as a measure to compare different grids. This is used in this context to determine suitable parameters for the rulebase system, but can be used to assess different grids as well.

The algorithm for spatial disaggregation and regridding is presented in Chap. 9. It constitutes a novel approach for a commonly recurring problem in spatial data processing; particularly novel is involving proxy data to improve the results. The connection to the data is determined automatically by trying different parameters (and parameter ranges) and assessing the similarity of parameter values with the ideal values in a training set. The method was presented and its disaggregation and regridding capabilities were tested with various examples in Sects. 11.2 and 11.3.

To perform experiments that investigate the behaviour and properties of the developed method, a few hurdles needed to be taken. These are explained in Chap. 10. This less concerns scientific research, but rather more practical solutions or algorithms that were developed in the context of the regridding algorithm yet may have other applications. As the developed method greatly depends on parameter values that involve geometries, calculations of geometries need to be performed accurately. The previsions in the framework to deal with robustness errors proved insufficient for this application, justifying the development of a framework for geometry calculations that were better suited for the problem (Sect. 10.2). For this, an approach was developed that compares the size of result of geometry calculations against the relative size of the arguments and decides whether or not this should be treated as is an erroneous case. This way, the method prevents false positives in e.g. an intersection test. It however does not catch false negatives in an intersection test, but for our application the presence of false positives proved far more problematic than the absence of false negatives. This is mainly the result of when/how intersections are used in the implementation rather than an issue with the regridding algorithm. A second implementation aspect concerns the determination of the ranges for the variables in the rulebase, presented in Sect. 10.3. The idea put forward in Chap. 5 requires most possible ranges to be determined for each data pair, but does not specify how to do this. Specifically, for the spatial problem considered here, two distinctively different ways of computing an optimal range for a data pair are presented in Sect. 10.3: local and estimated. The former considers value of a limited area around the cell of interest whereas the latter calculates upper and lower limit for the cell using specific formulas. For both, multiple possibilities and implementations are presented.

12.2 Future Directions for Research

The problem of spatial disaggregation and regridding spawned several new developments in different fields, but mainly in fuzzy rulebase systems. While the presented developments allow for the use of a fuzzy inference system to perform the spatial disaggregation or the regridding, further advancements would improve the methodology.

A first aspect is in the evaluation of the rulebase. In typical rulebase systems, all variables in a rule are considered equal and no weighting of the variables is possible. As the experiments that involve multiple proxy data have shown, not all proxy data is equally good or suitable. While data of lower quality can still contribute, it should be possible to lower the weight of this contribution to benefit data of higher quality. In addition, these weights should not be determined once for all rulebase applications but each time the rulebase is applied: the experiments have shown that some data may be better suited in one location than in another location. Allowing for a weighted evaluation of fuzzy rulebases is not only of importance for the spatial data problem, but would greatly improve the applicability of fuzzy rulebase systems. As such, this is one of the first directions in which the research continues.

Also related to the rulebase is the aspect of defuzzification. The currently developed method for defuzzification under constraints is an extension of the Mean of max method with the constraint that the sum of the defuzzified values is a known crisp value. Research in the extension of other defuzzification methods is interesting, as mean of max is quite limiting due to the fact that it does not consider the shape of the fuzzy set. The extension of constraint defuzzification to other defuzzifiers increases the applicability and allows for more appropriate defuzzification. The constraint that the sum of all values is a known crisp value is a simple constraint; future research is also aimed at considering other constraints imposed on the defuzzified values.

The following directions of research are closely connected to the developed algorithm. First, more research is needed on how to allow an expert to express the connection between proxy data and input data. At present, the system exhaustively tries all parameters to identify which parameters are suitable for the proxy data. An expert can easily say if the values in a proxy grid are proportional or inverse proportional to the input values. However, parameters are more complicated and contain many calculations with varying degrees of freedom, requiring the answer to questions such as *how is the minimum possible value calculated?* or *if a local range is used, how do we define the region in which is it considered?*. These questions have a big impact on the parameter but are difficult for an expert to determine. Ideally, their would be a mechanism that supports an expert in finding suitable parameters; helping the expert in answering such questions through e.g. a feedback loop on the training data.

A second aspect in the further development of the method is the automatic generation of a training set, as presented in Sect. 10.4. The current implementation only works if the proxy data is either proportional. The ability to generate a training set is very powerful as suitable training sets may be scarce due to the fact that the grid layouts have to match. Real world data can have many different connections, and allowing the generation of a training set that mimics this will improve ability of the system to correctly identify more data sets as suitable proxy data and also to better identify the most appropriate parameters. This is connected to the previous aspect, as most likely expert input will be needed to convey the system which parameter connections it should consider.

Closely connected to this is the final aspect: research in detecting the equivalence of parameters. Two parameters can have slightly different calculations but can have a very similar behaviour, in which case there is no knowledge to be gained by considering both of them in the rulebase. This aspect was the main reason for forcing the experiments that involve multiple proxy data to consider a parameter from each of the proxy data sets: the system otherwise would select very similar parameters that relate to one set and ignore the other datasets. The detection of such equivalence is not straightforward, but would allow the system to much faster dismiss a number of candidate parameters. This in turn would increase the speed of the training phase.

Moving away from theoretical research is the need for practical work to make the methodology more applicable. This not only means further developing the implementation in order to cope with more commonly used gridded formats, but also research in increasing the performance. Improvements are possible in the implementation by considering parallelism and even distributed computing, as the algorithm is can be parallelized. While the problem is not a real time problem as such, one could envision a real time application where the regridded dataset needs to be determined or modified on the fly based on incoming data; an agent-based approach may even be considered in this case. Performance can also be improved by caching values, but the size of the dataset in combination with the available hardware may limit how much data can be cached, resulting in an interesting balancing act between memory usage and cpu usage.

12.3 Other Application Fields

This algorithm was developed with the example of regridding air pollution data in mind, but can be applied more general. Currently, an application relating to grids that hold temperature data is considered. Due to the different nature of the data (not only is temperature not additive unlike pollution amounts, but negative values can occur and offset the average), different parameters and operations need to be developed and investigated.

The concept of using an artificial intelligent system for spatial disaggregation and regridding can also be applied for other purposes. In particular, an application of data fusion is being investigated. In data fusion, the goal is to combine the knowledge of multiple data sets into a single data set. This is to some extent similar to how multiple proxy data are used in the developed method in order to determine a value for a cell: the rulebase outputs a fuzzy set that is an estimate for the final value. The difference is that the range of the final value in the case of spatial disaggregation is determined by overlapping input cell, whereas in data fusion such a value would have to be calculated. More research is therefor necessary in determining how an appropriate value range can be determined in order to perform a data fusion.

A different application, which to some extent resembles data fusion is the identification of locations that meets certain criteria. In this case, the proxy data contains data that either positively or negatively affects the appreciation of a location. This idea has been envisioned in an archaeological context, where archaeologists are interested in finding new locations for surveys. For this, they would like to combine the knowledge on how a civilization established settlements, e.g. close to water, near arable land, not on a steep incline, etc. and determine potential locations for settlements. Such datasets can be considered as proxy data, with the conditions *close* or *far* determining the parameters that would be suitable in the rulebase; this problem then connects with the problem of involving expert knowledge as mentioned earlier. Alternatively, this problem could also be considered by first converting each of the datasets individually to a heatmap, a spatial map that identifies good locations based on a single criterion and then use the rulebase system to fuse the different heatmaps into a single aggregated heatmap. While this is again data fusion, the difference is that the range of the output is less of an issue in this application, as long as the values can be interpreted (e.g. a higher value equal a better location).

The developed methodology for spatial disaggregation and regridding not only provides for an algorithm to handle these specific spatial problems, but also resulted in a blueprint for solving other spatial data problems.

18. Jenks, G.F.: International yearbook of cartography: 1967. In: Frenzel, K. (ed.) The Data Model Concept in Statistical Mapping, vol. 7. George Philip (1967)
19. jFuzzyLogic: jfuzzylogic. http://jfuzzylogic.sourceforge.net/. Accessed 15 Jan 2018
20. Jonas, M., Marland, G., Krey, V., Wagner, F., Nahorski, Z.: Uncertainty in an emissions-constrained world. Clim. Change **124**(3), 459–476 (2014). https://doi.org/10.1007/s10584-014-1103-6
21. JTS: JTS Topology Suite. https://sourceforge.net/projects/jts-topo-suite/. Accessed Jan 15 2018; Current latest version is 1.14
22. JUMP: JUMP Pilot Project. https://sourceforge.net/projects/jump-pilot/. Accessed 15 Jan 2018; Current latest version is 2.18.15
23. Klir, G.J., Yuan, B.: Fuzzy Sets and Fuzzy Logic: Theory and Applications. Prentice Hall, New Jersey (1995)
24. Mamdani, E., Assilian, S.: An experiment in linguistic synthesis with a fuzzy logic controller. Int. J. Man-Mach. Stud. **7**(1), 1–13 (1975). https://doi.org/10.1016/S0020-7373(75)80002-2
25. Martinez-Urtaza, J., Bowers, J.C., Trinanes, J., DePaola, A.: Climate anomalies and the increasing risk of vibrio parahaemolyticus and vibrio vulnificus illnesses. Food Res. Int. **43**(7), 1780–1790 (2010). https://doi.org/10.1016/j.foodres.2010.04.001. http://www.sciencedirect.com/science/article/pii/S0963996910000980. Climate Change and Food Science
26. Mendel, J.M.: Uncertain Rule-based Fuzzy Logic Systems: Introduction and New Directions. Prentice Hall (2001)
27. Mugglin, A., Carlin, B., Zhu, L., Conlon, E.: Bayesian areal interpolation, estimation, and smoothing: an inferential approach for geographic information systems. Environ. Plan. A **31**(8), 1337–1352 (1999)
28. Mugglin, A.S., Carlin, B.P., Gelfand, A.E.: Fully model-based approaches for spatially misaligned data. J. Am. Stat. Assoc. **95**(451), 877–887 (2000)
29. Murata, T., Ishibuchi, H.: Adjusting membership functions of fuzzy classification rules by genetic algorithms. In: Proceedings of 1995 IEEE International Conference on Fuzzy Systems, vol. 4, pp. 1819–1824 (1995). https://doi.org/10.1109/FUZZY.1995.409928
30. National Research Council of Canada: FuzzyJ Toolkit. https://github.com/rorchard/FuzzyJ/. Accessed 15 Jan 2018; Current latest version is 2.0
31. Nomura, H., Hayashi, I., Wakami, N.: A self-tuning method of fuzzy reasoning by genetic algorithm. In: Proceedings of the 1992 International Fuzzy Systems and Intelligent Control Conference, pp. 236–245 (1992)
32. OpenGeoSpatial: Opengeospatial. http://www.opengeospatial.org/docs/is. Accessed 15 Mar 2018
33. Rafaj, P., Amann, M., Siri, J., Wuester, H.: Changes in European greenhouse gas and air pollutant emissions 1960–2010: decomposition of determining factors. Clim. Change **124**(3), 477–504 (2014). https://doi.org/10.1007/s10584-013-0826-0
34. Rigaux, P., Scholl, M., Voisard, A.: Spatial Databases with Applications to GIS. Morgan Kaufman Publishers (2002)
35. Shekhar, S., Chawla, S.: Spatial Databases: A Tour. Pearson Educations (2003)
36. Shewchuk, J.R.: Triangle: engineering a 2D quality mesh generator and delaunay triangulator. In: First Workshop on Applied Computational Geometry, pp. 124–133. Association for Computing Machinery, Philadelphia, Pennsylvania (1996)
37. Sugeno, M.: Industrial Applications of Fuzzy Control. Elsevier Science Inc., New York (1985)
38. Tobler, W.R.: Smooth pycnophylactic interpolation for geographic regions. J. Am. Stat. Assoc. **74**(367), 519–536 (1979)
39. Tomlin, C.: Special issue landscape planning: expanding the tool kit map algebra: one perspective. Landsc. Urban Plann. **30**(1), 3–12 (1994). https://doi.org/10.1016/0169-2046(94)90063-9
40. Van Leekwijck, W., Kerre, E.E.: Defuzzification: criteria and classification. Fuzzy Sets Syst. **108**, 159–178 (1999)
41. Verstraete, J.: Dealing with rounding errors in geometry processing. In: Flexible Query Answering Systems 2015–Proceedings of the 11th International Conference FQAS 2015, Cracow,

Poland, 26–28 October 2015, pp. 417–428 (2015). https://doi.org/10.1007/978-3-319-26154-6_32

42. Verstraete, J.: The spatial disaggregation problem: simulating reasoning using a fuzzy inference system. IEEE Trans. Fuzzy Syst. **25**(3), 627–641 (2016). https://doi.org/10.1109/TFUZZ.2016.2567452

43. Verstraete, J.: Fuzzy quality assessment of gridded approximations. Appl. Soft Comput. **55**, 319–330 (2017). https://doi.org/10.1016/j.asoc.2017.01.051. http://www.sciencedirect.com/science/article/pii/S1568494617300662

44. Volker, W., Fritsch, D.: Matching spatial datasets: a statistical approach. Int. J. Geograph. Inf. Sci. **13**(5), 445–473 (1999)

45. Wang, L.X., Mendel, J.M.: Generating fuzzy rules by learning from examples. IEEE Trans. Syst. Man Cybern. **22**(6), 1414–1427 (1992)

46. WikiPedia: Jenks natural breaks optimization. https://en.wikipedia.org/wiki/Jenks_natural_breaks_optimization. Accessed 17 Jan 2018

47. Yager, R.R., Filev, D.P.: Constrained defuzzification. In: Proceedings of the 5th IFSA Congress, pp. 1167–1170 (1993)

48. Yager, R.R., Filev, D.P.: Defuzzification with constraints. In: Fuzzy Logic and its Applications to Engineering, Information Sciences, and Intelligent Systems–Theory and Decision Library, vol. 16, pp. 157–166 (1996)

49. Zadeh, L.A.: Fuzzy sets. Inf. Control **8**, 338–353 (1965)

50. Zimmerman, H.J.: Practical Applications of Fuzzy Technologies. Kluwer Academic Publishers (1999)

Printed in the United States
By Bookmasters